Improving Representation in Clinical Trials and Research

BUILDING RESEARCH EQUITY FOR
WOMEN AND UNDERREPRESENTED GROUPS

Kirsten Bibbins-Domingo and Alex Helman, *Editors*

Committee on Improving the Representation of Women and Underrepresented Minorities in Clinical Trials and Research

Committee on Women in Science, Engineering, and Medicine

Policy and Global Affairs

A Consensus Study Report of

The National Academies of
SCIENCES • ENGINEERING • MEDICINE

THE NATIONAL ACADEMIES PRESS
Washington, DC
www.nap.edu

THE NATIONAL ACADEMIES PRESS   **500 Fifth Street, NW**   **Washington, DC 20001**

This activity was supported by contracts between the National Academy of Sciences and the National Institutes of Health (HHSN263201800029I/75N98020F00023). Any opinions, findings, conclusions, or recommendations expressed in this publication do not necessarily reflect the views of any organization or agency that provided support for the project.

International Standard Book Number-13: 978-0-309-27820-1
International Standard Book Number-10: 0-309-27820-1
Digital Object Identifier: https://doi.org/10.17226/26479
Library of Congress Control Number: 2022940407

Additional copies of this publication are available from the National Academies Press, 500 Fifth Street, NW, Keck 360, Washington, DC 20001; (800) 624-6242 or (202) 334-3313; http://www.nap.edu.

Copyright 2022 by the National Academy of Sciences. All rights reserved.

Printed in the United States of America

Suggested citation: National Academies of Sciences, Engineering, and Medicine. 2022. *Improving Representation in Clinical Trials and Research: Building Research Equity for Women and Underrepresented Groups*. Washington, DC: The National Academies Press. https://doi.org/10.17226/26479.

The National Academies of
SCIENCES · ENGINEERING · MEDICINE

The **National Academy of Sciences** was established in 1863 by an Act of Congress, signed by President Lincoln, as a private, nongovernmental institution to advise the nation on issues related to science and technology. Members are elected by their peers for outstanding contributions to research. Dr. Marcia McNutt is president.

The **National Academy of Engineering** was established in 1964 under the charter of the National Academy of Sciences to bring the practices of engineering to advising the nation. Members are elected by their peers for extraordinary contributions to engineering. Dr. John L. Anderson is president.

The **National Academy of Medicine** (formerly the Institute of Medicine) was established in 1970 under the charter of the National Academy of Sciences to advise the nation on medical and health issues. Members are elected by their peers for distinguished contributions to medicine and health. Dr. Victor J. Dzau is president.

The three Academies work together as the **National Academies of Sciences, Engineering, and Medicine** to provide independent, objective analysis and advice to the nation and conduct other activities to solve complex problems and inform public policy decisions. The National Academies also encourage education and research, recognize outstanding contributions to knowledge, and increase public understanding in matters of science, engineering, and medicine.

Learn more about the National Academies of Sciences, Engineering, and Medicine at **www.nationalacademies.org**.

The National Academies of
SCIENCES • ENGINEERING • MEDICINE

Consensus Study Reports published by the National Academies of Sciences, Engineering, and Medicine document the evidence-based consensus on the study's statement of task by an authoring committee of experts. Reports typically include findings, conclusions, and recommendations based on information gathered by the committee and the committee's deliberations. Each report has been subjected to a rigorous and independent peer-review process and it represents the position of the National Academies on the statement of task.

Proceedings published by the National Academies of Sciences, Engineering, and Medicine chronicle the presentations and discussions at a workshop, symposium, or other event convened by the National Academies. The statements and opinions contained in proceedings are those of the participants and are not endorsed by other participants, the planning committee, or the National Academies.

For information about other products and activities of the National Academies, please visit www.nationalacademies.org/about/whatwedo.

COMMITTEE ON IMPROVING THE REPRESENTATION OF WOMEN AND UNDERREPRESENTED MINORITIES IN CLINICAL TRIALS AND RESEARCH

KIRSTEN BIBBINS-DOMINGO, M.D., Ph.D., M.A.S. (*Chair*) (NAM),[1] Professor and Chair of the Department of Epidemiology and Biostatistics and Lee Goldman, MD Endowed Chair and Professor of Medicine, University of California, San Francisco; Inaugural Vice Dean for Population Health and Health Equity, UCSF School of Medicine

MARCELLA ALSAN, M.D., Ph.D., M.P.H., Professor of Public Policy, Harvard Kennedy School; Co-Director of the Health Care Delivery Initiative Abdul Latif Jameel Poverty Action Lab, Massachusetts Institute of Technology

ARLEEN BROWN, M.D., Ph.D., Professor of Medicine, University of California, Los Angeles; Chief of General Internal Medicine and Health Services Research, Olive View-UCLA Medical Center

GLORIA CORONADO, Ph.D., Epidemiologist and Mitch Greenlick Endowed Distinguished Investigator in Health Disparities Research, Kaiser Permanente Center for Health Research

CARLOS DEL RIO, M.D. (NAM), Distinguished Professor of Medicine, Emory University School of Medicine; Professor of Epidemiology and Global Health, Rollins School of Public Health, Emory University; Executive Associate Dean for Emory University School of Medicine, Grady Health System

XINQI DONG, M.D., M.P.H., Director of the Institute for Health, Health Care Policy, and Aging Research (IFH) and Inaugural Henry Rutgers Professor of Population Health Sciences, Rutgers University

DANA GOLDMAN, Ph.D. (NAM), Dean of the Sol Price School of Public Policy, C. Erwin and Ione L. Piper Chair, and Leonard D. Schaeffer Director's Chair, Schaeffer Center for Health Policy & Economics, University of Southern California

SHARON K. INOUYE, M.D., M.P.H. (NAM), Professor of Medicine, Harvard Medical School; Milton and Shirley F. Levy Family Chair, and Director, Aging Brain Center, Marcus Institute for Aging Research, Hebrew SeniorLife

JONATHAN JACKSON, Ph.D.,[2] Executive Director of Community Access, Recruitment, and Engagement (CARE) Research Center, Massachusetts General Hospital and Harvard Medical School

---

[1] Designates membership in the National Academy of Sciences (NAS), National Academy of Engineering (NAE), or National Academy of Medicine (NAM).

[2] Dr. Jonathan Jackson resigned from the Committee on Improving Representation of Women and Underrepresented Minorities in Clinical Trials and Research, effective June 2, 2021.

AMELIA KNOPF, Ph.D., M.P.H., Assistant Professor of Nursing, Indiana University
EDITH A. PEREZ, M.D., Chief Medical Officer of Bolt Therapeutics, Inc.; Professor of Medicine, Mayo Clinic; Director of the Mayo Clinic Breast Cancer Translational Genomics Program
PHYLLIS PETTIT NASSI, B.S., M.S.W., Associate Director for Research and Science, Special Populations, American Indian Program, Huntsman Cancer Institute, University of Utah
JASON RESENDEZ, B.A., President and CEO, National Alliance for Caregiving
SUSAN SCHAEFFER, B.F.A., Founder, President, and CEO of The Patients' Academy for Research Advocacy

Study Staff

ALEX HELMAN, Ph.D., Study Director and Program Officer, Committee on Women in Science, Engineering, and Medicine
ASHLEY BEAR, Ph.D., Director, Committee on Women in Science, Engineering, and Medicine
LAURA AIUPPA, M.S., Senior Program Officer, Board on Health Care Services
AUSTEN APPLEGATE, Research Associate, Committee on Women in Science, Engineering, and Medicine
JOHN VERAS, Senior Program Assistant, Committee on Women in Science, Engineering, and Medicine (August 2020 to February 2022)
ABIGAIL HARLESS, Senior Program Assistant, Committee on Women in Science, Engineering, and Medicine (from February 2022)
MOLLIE MARR, Ph.D., Mirzayan Fellow, Committee on Women in Science, Engineering, and Medicine
ANNE MARIE HOUPPERT, MSLS, Senior Librarian

Consultants

FRANCHESCA ARIAS, Ph.D., Instructor in Neurology and Assistant Scientist, Harvard Medical School
ADAM BRESS, PharmD., M.S., National Academy of Medicine Fellow in Pharmacy
BRANDON BROWN, Ph.D., M.P.H., Associate Professor, University of California, Riverside
FARAH ACHER KAIKSOW, M.D., M.P.P., Assistant Professor of Medicine, Harvard Medical School
AISHWARYA BHATTACHARYA, M.P.H., Student Research Assistant, Yale School of Public Health

LAURA BOTHWELL, Ph.D., M.Phil., M.A., Associate Research Scientist, Yale School of Public Health
JOCELYN CARTER, M.D., M.P.H., Professor of Medicine, Harvard Medical School
AMBER DATTA, M.P.H., Student Research Assistant, Yale School of Public Health
JAKUB HLÁVKA, Ph.D., Fellow, USC Schaeffer Center for Health Policy & Economics; Research Assistant Professor, Health Policy and Management, USC Price School of Public Policy; Research Fellow, USC Center for Risk and Economic Analysis of Terrorism Events (CREATE)
AARON KESSELHEIM, M.D., J.D., M.P.H., Professor of Medicine, Harvard Medical School
AMY J. H. KIND, M.D., Ph.D., Professor of Medicine, University of Wisconsin School of Medicine and Public Health
NIROOP RAJASHEKAR, B.S., Student Research Assistant, Yale School of Medicine
NICOLE ROGUS-PULIA, Ph.D., M.A., Assistant Professor of Medicine and Surgery, University of Wisconsin School of Medicine and Public Health
BRYAN TYSINGER, Ph.D., Director of Health Policy Microsimulation, USC Schaeffer Center for Health Policy & Economics; Fellow, USC Schaeffer Center; Research Assistant Professor, USC Price School of Public Policy
LESLIE WANG, M.S., Student Research Assistant, Yale School of Medicine

Acknowledgments

This Consensus Study Report was reviewed in draft form by individuals chosen for their diverse perspectives and technical expertise. The purpose of this independent review is to provide candid and critical comments that will assist the National Academies of Sciences, Engineering, and Medicine in making each published report as sound as possible and to ensure that it meets the institutional standards for quality, objectivity, evidence, and responsiveness to the study charge. The review comments and draft manuscript remain confidential to protect the integrity of the deliberative process.

We thank the following individuals for their review of this report:

Margarita Alegria, Harvard University; Susan Ellenberg, University of Pennsylvania; Ayana Elliott, Gilead Sciences, Inc.; Celia Fisher, Fordham University; Sara Goldkind, Independent Consultant; Gina Green-Harris, University of Wisconsin; Jose Pagán, New York University; Eric Rubin, New England Journal of Medicine; William Schpero, Cornell University; Joshua Sharfstein, Johns Hopkins University; Jonathan Skinner, Dartmouth College; Stephanie Studenski, University of Pittsburgh; and Consuelo Wilkins, Vanderbilt University.

Although the reviewers listed above provided many constructive comments and suggestions, they were not asked to endorse the conclusions or recommendations of this report nor did they see the final draft before its release. The review of this report was overseen by Thomas LaVeist, Tulane University, and Eve Higginbotham, University of Pennsylvania. They were responsible for making certain that an independent examination of this report was carried out in accordance with the standards of the National Academies and that all review comments were carefully considered. Responsibility for the final content rests entirely with the authoring committee and the National Academies.

Preface

The year the issues in this report, *Improving Representation in Clinical Trials and Research*, became urgent for me was 2017. I was then chair of the U.S. Preventive Services Task Force (USPSTF)—an independent body charged with reviewing the scientific literature to generate evidence-based guidelines on the use of clinical preventive services. USPSTF guidelines are widely disseminated, and their audience includes patients, clinicians, and policy makers alike. During my tenure, we had issued recommendations on preventing diabetes and common cancers such as breast, colorectal, lung, and prostate that are responsible for considerable morbidity and mortality in the United States, as well as being important contributors to health disparities. Reaching patients and frontline clinicians directly was particularly compelling given the exceptionally strong evidence that clinical interventions work in preventing these diseases and because the Affordable Care Act had ensured that interventions for which evidence was clear would be covered by commercial insurers.

In my formal talks and informal discussions with lay and professional stakeholders, I inevitably encountered a similar pattern of questions:

> How confident are you that these recommendations and the evidence on which they are based apply to me and to patients like me?

> You are recommending screening for diabetes in those who are overweight or obese, but my Asian patients seem to develop diabetes at lower BMI, what about them?

> What about my Latino patients who are developing diabetes at younger ages or my Black patients who are developing colorectal cancer at younger ages— shouldn't we start screening earlier?

Black women get breast cancer at the same rate as others, but are more likely to die—should we screen differently?

My recurring response was, "Unfortunately, we just don't have the studies in these populations that allow us to say with certainty whether or how to adapt our prevention guidelines." While it is true, this answer rang hollow. As a physician caring for patients in an urban safety-net setting and wanting to provide the best evidence-based preventive care, these were my questions as well. Inevitably in these sessions, I would spend as much time on the science as I devoted to reinforcing with patients why they should still trust these guidelines and the process, despite the unrepresentative populations in the evidence base. With clinicians, we discussed how we might adapt the guidelines to the needs of our patient populations, what kind of evidence would be necessary, and how we might advocate together to ensure that coverage was preserved.

The year these issues became personal for me was also 2017. This was the year my father lost his battle with prostate cancer and another very close family member received a new diagnosis of this same disease. Prostate cancer is the most common cancer in men in the United States; its incidence in Black men (like the two in my family) is at least 75 percent higher than men of other races and ethnicities. My father was fortunate to have received care from outstanding physicians and to have had access to clinical trials as his disease advanced. He was a career Army officer, a veteran, and a strong supporter of science and medicine. He had even served as a lay reviewer for federal funding of prostate cancer research. As my father's journey with prostate cancer ended and another family member's began coincident with my work on the USPSTF, the stark absence of representation of Black men in prostate cancer research became acutely distressing. Black men constitute 13.4 percent of the U.S. population, have a higher prostate cancer incidence, and die at double the rate of other men in the United States. Yet the screening trials from which the USPSTF derived evidence for prevention included less than 5 percent Black men, and the number in late-stage treatment trials was recently reported at 2.4 percent.

I am grateful to have worked with the excellent members of this consensus committee. All generously volunteered their time and expertise over the past 18 months to develop an approach to this report and to crafting recommendations on improving representation in clinical trials and clinical research. I am grateful to the many experts who shared their knowledge of the complex clinical research landscape in our public meetings and to the outstanding teams that created our four commissioned papers. And I am particularly grateful to the National Academies of Sciences, Engineering, and Medicine staff, led by Dr. Alex Helman, who adroitly guided this complex work during a global pandemic over Zoom, by phone, and via email. Most of all, I am grateful that across the different perspectives and points of view on the nuances of these issues, all who were involved shared an understanding of their importance and producing a report with findings and actionable recommendations that would improve outcomes.

I hope that you will read this report in its entirety, through to the epilogue, where the committee envisions a better world for clinical research. I hope you will read with the intention to work to implement our findings and recommendations in whatever part of the clinical research ecosystem you influence. Whether you are motivated by the goal of producing the highest quality science, by pursuit of fairness and equity in how science might translate into better health for our patients, or by the enormous economic toll of health disparities in the United States, I hope you embrace the urgency of improving representation and inclusion in clinical research.

Kirsten Bibbins-Domingo, *Chair*

Contents

SUMMARY   1

1   THE COMMITTEE'S TASK   15
Committee Task and Approach, 17
Definition of Terms, 20
Organization of the Report, 21

2   WHY DIVERSE REPRESENTATION IN CLINICAL
RESEARCH MATTERS AND THE CURRENT STATE OF
REPRESENTATION WITHIN THE CLINICAL RESEARCH
ECOSYSTEM   23
Lack of Representation in Clinical Research Threatens the
Overarching Goals of Clinical Research, 23
Clinical Trials Production Process, 33
Current Status of Clinical Trial and Clinical Research Participation:
    Little Change Over Time, 35

3   POLICIES TO IMPROVE CLINICAL TRIAL AND
RESEARCH DIVERSITY: HISTORY AND FUTURE
DIRECTIONS   47
Early History, 47
Modern Policies, 50
Special Populations, 70

4	**BARRIERS TO REPRESENTATION OF UNDERREPRESENTED AND EXCLUDED POPULATIONS IN CLINICAL RESEARCH** Individual and Community Factors, 75 Individual Research Studies, 82 Landscape for Research—Community and Policy Factors That Influence the Representativeness of Clinical Trials and Research, 91	75
5	**FACILITATORS OF SUCCESSFUL INCLUSION IN CLINICAL RESEARCH** Insights into Effective Facilitators and Strategies for Inclusion, 108 Conclusions, 122	107
6	**RECOMMENDATIONS FOR IMPROVING REPRESENTATION IN CLINICAL TRIALS AND CLINICAL RESEARCH** Conclusions, 127 Recommendations, 129	127

EPILOGUE: ENVISIONING A NEW FUTURE 135
The Science of Inclusion, 135
Embracing Justice, 136

REFERENCES 139

APPENDIXES
A Quantifying the Potential Health and Economic Impacts of Increased Trial Diversity 169
B Key Trends in Demographic Diversity in Clinical Trials 191
C Improving Representativeness in Clinical Trials and Research: Facilitators to Recruitment and Retention of Underrepresented Groups 213

Boxes, Figures, and Tables

BOXES

1-1 Congressional Mandate, 18
1-2 Statement of Task, 19

2-1 Example: Adjustment of Dosing for Warfarin, 25
2-2 Economic Cost of Lack of Representation in Clinical Trials and Research, 27
2-3 Future Elderly Model, 28
2-4 Task Force on Research Specific to Pregnant Women and Lactating Women Report Recommendations, 42

4-1 Community Health Applied Research Network, 98

5-1 Federal Support of Early-Career Researchers Can Affect Access to Opportunities, 116
5-2 Promising Practices for Supporting a More Diverse and Equitable Medical Workforce, 117
5-3 Patient-Centered Outcomes Research Institute: Supporting Engagement, 121
5-4 Case Study: The Wisconsin Alzheimer's Institute, Regional Milwaukee Office Community Engagement, 123

FIGURES

2-1 Overview of the clinical trial ecosystem, 34
2-2 Average percent of women in trials by year of FDA approval and therapeutic area (n = 287), 39
2-3 Participation of women in clinical trials supported by NIH institutes (top 10 institutes/centers by 2018 enrollment), 39

5-1 Improving diversity in enrollment, 119

B-1 Average % of females in trials by year of FDA approval and therapeutic area (n = 287), 197
B-2 Mean % of females by year of FDA approval (non-gender-specific trials only, n = 255), 197
B-3 Average % of white patients in trials by year of FDA approval and therapeutic area (n = 287), 198
B-4 Average % of patients over 65 in trials by year of FDA approval and therapeutic area (n = 287), 199
B-5 Participation of females in clinical trials supported by NIH institutes (top 10 institutes/centers by 2018 enrollment), 200
B-6 Share of white participants in clinical trials by NIH institutes (top 10 institutes/centers by 2018 enrollment), 201
B-7 Share of African American/Black participants in clinical trials by NIH institutes (top 10 institutes/centers by 2018 enrollment), 202
B-8 Share of Asian participants in clinical Trials by NIH institutes (top 10 institutes/centers by 2018 enrollment), 203
B-9 Share of Hispanic participants in clinical trials by NIH institutes (top 10 institutes/centers by 2018 enrollment), 203
B-10a Availability of results among all trials, by primary completion year, 205
B-10b Availability of results among Phase 3 trials, by primary completion year, 205

C-1 Flow chart illustrating process for identification of trials, 222

TABLES

2-1 Adjusted Relative Risks for Key Parameters of Interest with 95% Confidence Intervals, 29

3-1 2020 Life Expectancy at Birth, 72

4-1 How Specific Community-Engagement Opportunities Can Benefit Research Organizations and Communities, 94

5-1 Strategies to Achieve Representation in Clinical Research by Theme, 124

A-1 1998–2018 Health and Retirement Study Sample Characteristics, 172
A-2 Adjusted Relative Risks for Key Parameters of Interest, 173
A-3 Baseline Characteristics at Simulation Start, 174
A-4 Life Years, Disability-free Life Years, and Remaining Work Years for Diabetes Scenario, 176
A-5 Life Years, Disability-free Life Years, and Remaining Work Years for Heart Disease Scenario, 176
A-6 Life Years, Disability-free Life Years, and Remaining Work Years for Hypertension Scenario, 177
A-7 Aggregate Value of Diabetes Scenario, 178
A-8 Aggregate Value of Heart Disease Scenario, 180
A-9 Aggregate Value of Hypertension Scenario, 181
A-10 Population Value for Scenarios through 2050, 182
A-11 Diabetes, 183
A-12 Heart Disease, 185
A-13 Hypertension, 187

B-1 Demographics of Participants in Trials Supported by NIH Centers and Institutes, 201

C-1 Number of Trials in Each Disease Category by U.S. Census Geographic Regions and Divisions, 221
C-2a Starting with Intention and Agency to Achieve Representativeness, 235
C-2b Establishing a Foundation of Trust with Participants and Community, 240
C-2c Anticipating and Removing Barriers to Study Participation, 246
C-2d Adopting a Flexible Approach to Recruitment and Data Collection, 248
C-2e Building a Robust Network by Identifying All Relevant Stakeholders, 250
C-2f Navigating Scientific, Professional Peer, and Social Expectations, 252
C-2g Optimizing Study Team to Ensure Alignment with Research Goals, 254
C-2h Attaining Resources and Support to Accomplish a Representativeness, 256

Summary

The United States has long made substantial investments in clinical research with the goal of improving the health and well-being of our nation. There is no doubt that these investments have contributed significantly to treating and preventing disease and extending human life. Nevertheless, clinical research faces a critical shortcoming. Currently, large swaths of the U.S. population, and those that often face the greatest health challenges, are less able to benefit from these discoveries because they are not adequately represented in clinical research studies.

In the past three decades, diversity in clinical trials has become an important policy priority, advanced by federal agency offices such as the National Institutes of Health (NIH) Office of Research on Women's Health, the Food and Drug Administration (FDA) Office of Women's Health, the Society for Women's Health Research, and the FDA Office of Minority Health. While progress has been made on some fronts, particularly with representation of white women in clinical trials and clinical research, progress has largely stalled on participation of racial and ethnic minority population groups. Additionally, older adults, pregnant and lactating individuals, LGBTQIA+ populations, and persons with disabilities remain underrepresented and even excluded from clinical trials and clinical research.[1] An equitable clinical research enterprise would include trials and studies that match the demographics of the disease burden under study. However, we remain far from achieving this goal.

---

[1] Throughout this report, LGBTQIA+ is used as an inclusive term for the various gender identities and sexual orientations, including lesbian, gay, bisexual, transgender, questioning, queer, intersex, asexual, and pansexual.

By failing to achieve a more diverse clinical trial and clinical research enterprise, the nation suffers serious costs and consequences, including the following:

1. **Lack of representation compromises generalizability of clinical research findings to the whole U.S. population.** Women, pregnant people, children, older adults, and racial and ethnic minority population groups can have distinct disease presentations or health circumstances that affect how they will respond to an investigational drug or therapy. These variable therapeutic responses can result in the delivery of health care that is not always evidence based.
2. **Lack of representation costs hundreds of billions of dollars.** An economic analysis carried out by the committee, using the Future Elderly Model, demonstrates high financial and social costs, measured by life expectancy, disability-free life, and years in the labor force, in the hundreds of billions of dollars range (see Box 2-1). Given the assumption that better representation in clinical trials would reduce health disparities by even a modest amount, the analysis found that achieving diverse representation in research would be worth billions of dollars in savings to the United States.
3. **Lack of representation may hinder innovation and new discoveries.** Diversity in study participants allows for greater exploration of variation in the overall effectiveness of a particular intervention. Exploring "heterogeneity of treatment effects" may be necessary not only to understand variation that affects safety and effectiveness of an intervention in underrepresented and excluded populations but also to identify new biological processes that may, in turn, lead to new discoveries important for all populations.
4. **Lack of representation may compound low accrual that causes many trials to fail.** According to an analysis by GlobalData, low accrual was the cause for stopping 55 percent of all Phase I–IV clinical trials that were terminated, suspended, or discontinued during 2008–2017. Thus, increasing enrollment of underrepresented and excluded populations would help solve the leading cause of clinical trial failure.
5. **Lack of representation may lead to lack of access to effective medical interventions.** Approval and indications for new therapeutics are often restricted to the demographics of the populations included in the clinical studies. Lack of representation may therefore impede access to a specific therapeutic agent. Guideline-making bodies must synthesize various lines of evidence when making recommendations. The generalizability of these recommendations to all populations may be limited when the evidence base for a specific population does not exist. When these recommendations are tied to insurance coverage, these gaps may affect reimbursement of, and therefore access to, health care.

SUMMARY

6. **Lack of representation may undermine trust of the clinical research enterprise and the medical establishment.** For example, the lack of inclusion of pregnant people in the clinical trials of the SARS-CoV-2 vaccines led to lack of clarity on the use of these vaccines in pregnant people and may have contributed to vaccine hesitancy, even as subsequent observational data emerged showing the safety of vaccine use in pregnant individuals, as well as data on the importance of preventing COVID-19 infection during pregnancy. Efforts to create more representative and inclusive research environments may work to increase trust in science and medicine.
7. **Lack of representation compounds health disparities in the populations currently underrepresented and excluded in clinical trials and clinical research.** While achieving health equity and reducing health disparities requires far more than just equitable representation in clinical research, failure to achieve equity on this dimension leaves health disparities unaddressed and reinforces inequities.

STATUS OF CLINICAL TRIAL PARTICIPATION

Gaining a fully accurate status of the current participation of underrepresented populations in clinical trials and clinical research, and trends in participation over time, is very challenging due to insufficient data-reporting practices at a national level. Although reporting to ClinicalTrials.gov is required for ongoing studies, the committee found major inconsistencies in how data were reported in this national database. Further, NIH does not currently have longitudinal data available for clinical trial enrollment by disease type.

Working within these constraints, the committee commissioned an analysis to examine available data from the FDA and NIH, which found that women now represent greater than 50 percent of clinical trial participants in the United States, particularly for white women. However, pregnant and lactating individuals, sexual- and gender-minority populations, and racial and ethnic subgroups of women remain underrepresented in clinical trials. The analysis also revealed that the racial and ethnic diversity of clinical trials is largely stagnant, with little changes in diversity over time.

UNDERREPRESENTED AND EXCLUDED POPULATIONS ARE WILLING TO PARTICIPATE IN CLINICAL RESEARCH, IF ASKED

Due to well-documented historical and contemporary abuses against certain excluded and underrepresented populations in medical research, members of the research community often assume that a lack of willingness to participate in research is the major driver of poor representation of some populations in research. However, the evidence on this issue is clear: Asian, Black, Latinx Americans, and

American Indian/Alaska Native individuals are no less likely, and in some cases are more likely, to participate in research if they are asked. Distrust and mistrust are commonly assumed to be the reason underlying a lack of participation in clinical trials. While there is no doubt that the legacy of abuses in medical research is an important factor driving the lack of engagement of underrepresented and excluded populations with both health care and research, several studies have found that distrust and mistrust are not necessarily associated with a lack of willingness to participate in medical research. The evidence suggests that concerns of researchers about the willingness of underrepresented and excluded populations to participate in research due to distrust or mistrust in the medical establishment may misrepresent barriers to participation in research or are surmountable with effort from research teams, funders, and policy makers.

BARRIERS TO REPRESENTATION OF UNDERREPRESENTED AND EXCLUDED POPULATIONS IN CLINICAL RESEARCH

The committee found that the existing research system has served to reduce participation by a diverse population in clinical trials and clinical research through a range of factors, operating at multiple levels. Individual research studies, the institutions that conduct research, funders of studies, institutional review boards (IRBs), medical journals, and the broader landscape of national policies and practices that govern research can all contribute to barriers to inclusion of underrepresented and excluded populations in clinical research.

1. **Individual research studies.** At the level of an individual research study, the factors and problems that lead to the underrepresentation and exclusion of certain populations in clinical trials and research begin with and follow the life cycle of a project. Understanding and resolving underrepresentation and exclusion of these populations in research requires careful examination of almost every stage in the research process itself, including
    - the development of research questions;
    - the composition, training, and attitudes of the research team;
    - research site selection;
    - participant selection, including sampling and recruitment methods and inclusion and exclusion criteria;
    - study protocols, including informed consent processes and remuneration; and
    - development and inclusion of multilingual recruitment and consent documents.
2. **Institutional structures.** Medical institutions of different types face a range of structural barriers to inclusion in clinical trials. For example, although academic medical centers conduct 55 percent of the extramural medical research supported by the NIH, and operate 98 percent of

the nation's 41 comprehensive cancer centers as of 2019, sustainably and meaningfully engaging underrepresented and excluded populations often does not align with the traditional incentive structures for researchers at these institutions. Recruiting diverse population groups and properly engaging with community members, which is time-consuming and requires investments to build and sustain trust, are only minimally considered in promotion and tenure decisions at academic medical centers. And while community health centers serve a much more diverse community than academic medical centers, these institutions also face barriers to clinical trials and research recruitment, which, which include limited provider knowledge about available research opportunities and challenges with electronic health record (EHR) infrastructure that can limit providers' ability to query the EHR using study inclusion and exclusion criteria.
3. **Institutional review boards.** IRBs can also present barriers to diverse participation in clinical trials by limiting the types and amount of compensation given to research participants to avoid the impression of coercion or undue influence. However, limiting incentives may ultimately compromise beneficence and justice, two of the ethical principles for research with human subjects detailed in the *Belmont Report*.
4. **Research funders.** Research funders also have several roles and responsibilities that can influence the diversity of clinical trials. These include setting funding priorities, deciding which projects ultimately get funded, providing adequate funding to recruit and retain participants, requiring transparent reporting, and evaluating research outputs.
5. **Industry funders.** Most clinical trials are funded by industry, and these trials present barriers, including out-of-pocket costs for participants, which are often not discussed in the informed consent process, industry pressures to gather data quickly, and the selection of easy-to-recruit samples being incentivized. It should be noted that some of these barriers are not solely unique to industry-sponsored trials.
6. **Medical journals.** Peer-reviewed Medical journals serve as the gatekeepers to scientific advancements in clinical practice and health. Their editors yield great power for what is, and is not, published in their pages. Lack of representation on editorial boards and other journal leadership positions may contribute to biases in publication.

FACILITATORS TO SUCCESSFUL INCLUSION IN RESEARCH

There is substantial quantitative data demonstrating the size and scope of the problem of underrepresentation and exclusion of populations in research; however, there is a dearth of critical qualitative data about facilitators of successful inclusion in clinical research. This committee supplemented existing literature with commissioned research with 20 researchers who worked on trials that met criteria for diverse trial enrollment. From this research, eight major themes emerged, which provide insights into key facilitators to inclusion:

1. **Starting with intention and agency to achieve representativeness.** From goal setting to community partnering strategies, intentionality and planning are critical themes for overcoming the systemic barriers previously outlined to the inclusion of underrepresented and excluded populations in research. This intentionality applies to building relationships with community members, designing studies that seek to recruit these groups, considering barriers to access and the lived-realities of participants in the research design, and external factors, such as requirements from funding agencies.
2. **Establishing a foundation of trust with participants and the community at large.** Building and maintaining trust with both study participants and their larger communities is foundational to achieving equity in research. The development of trust requires a long-term commitment by principal investigators, study teams, and local institutions involved in the research. Building trust over time takes consistent engagement in the community beyond the confines of the study itself, developing meaningful relationships with study participants, and giving to the community without the expectation of anything in return.
3. **Anticipating and removing barriers to study participation.** Building rapport with study participants and attending to their needs is critical for making sure studies have broad accessibility. In addition, recognizing heterogeneity within cultural groups is key; a one-size-fits-all approach to developing protocols will not work.
4. **Adopting a flexible approach to recruitment and data collection.** Flexibility in recruitment techniques, data collection, and visit windows to adapt to study needs is critical to having diverse study enrollment and retention. These changes are more helpful when made with input from community representatives and other relevant stakeholders.
5. **Building a robust network by identifying all relevant stakeholders.** Research suggests that engaging in mapping to identify all the relevant stakeholders in a community can help study teams develop more equitable study designs and identify individuals and organizations that can help drive the recruitment and retention of diverse study participants. These stakeholders include caregivers, family members, friends, clinical providers and administrators, community advocates, peers, religious leaders, and political figures.
6. **Navigating scientific, professional peer, and societal expectations.** Efforts to promote representativeness, and decisions made to support these efforts, are not always embraced or supported by colleagues and

organizations responsible for making funding and/or budget decisions. It is helpful if funding agencies, as well as those responsible for approving proposals and distributing budgets, understand the challenges and costs associated with nontraditional research approaches to enhance inclusion.
7. **Optimizing the study team to ensure alignment with research goals.** Diverse study teams, including study leadership, are helpful to recruitment and to enhance congruence between research teams and potential participants. It also helps to retain staff over time for recruitment and retention success.
8. **Attaining resources and support to achieve representativeness.** The investment of time and money are necessary to successfully engage in the long-term strategies and relationship building needed to drive inclusion in studies. This includes expanded budgets for teams recruiting and retaining diverse participants, support to expand infrastructure for community organizations, and investments in community-based partnerships to reduce power differentials between researchers and participants.

CONCLUSIONS

The committee identified five overarching conclusions, based on a comprehensive analysis of the research, presented throughout the report, which serve to frame the consensus recommendations.

1. **Improving representation in clinical research is urgent.**
   The scientific necessity to improve research equity is urgent. The 2020 U.S. Census found that the number of people who identify as white has shrunk for the first time since a census started being taken in 1790, and despite the country becoming more diverse, the nation's health disparities persist. Without major advancements in the inclusion of underrepresented and excluded populations in health research, meaningful reductions in disparities in chronic diseases such as diabetes, cancer, and Alzheimer's remain unlikely. Purposeful and deliberate change is needed. As the United States becomes more diverse every day, failing to reach these growing communities will only prove more costly over time (see Chapter 2).
2. **Improving representation in clinical research requires investment.**
   Improving the representation of underrepresented and excluded populations in clinical trials and clinical research requires a substantial investment of time, money, and effort. Investment of time and resources are needed to build and restore trust with underrepresented and excluded communities. Building trust with local communities cannot be episodic or transactional and pursued only to meet the goals of specific studies; it requires sustained presence, commitment, and investment. Investments are also needed in the systems and technologies that reduce burdens to

participation by underrepresented and excluded populations, such as by adequately compensating participants financially for their time when participating in research and by investing resources in making participation more physically accessible, and by providing research materials that are culturally informed and multilingual. Lastly, we need to invest in creating a more diverse workforce that better reflects the diversity of our country. This not only has implications for study site personnel and their direct interactions with participants, but also influences the types of research questions that get asked, the types of research that get funded, and even the types of research that are published. To better address health disparities and ensure health equity for all, the U.S. workforce should look more like the nation (see Chapter 4).

3. **Improving representation requires transparency and accountability.**

Transparency and accountability throughout the entire research enterprise will be critical to driving change and must be present at all points in the research life cycle—from the questions being addressed, to ensuring the populations most affected by the health problems are engaged and considered in the design of the study, to recruitment and retention of study participants, to analysis and reporting of results. Individual investigators and research institutions on the front lines bear responsibility for transparency in reporting progress toward the goals of inclusion in research. Transparency and accountability must also be reinforced by the funding that agencies and industry sponsors have across their portfolios, that regulatory agencies have in their role governing the conduct of research as well as the approval and reimbursement of the drugs and devices that are often the final products of clinical research, and that journal editors and others that disseminate research have in communicating findings (see Chapters 3, 4, and 5).

4. **Improving representation in clinical research is the responsibility of everyone involved in the clinical research enterprise.**

The clinical research landscape is complex and involves multiple stakeholders—participants, communities, investigators, IRBs, industry sponsors, institutions, funders, regulators, journals, and policy makers. Each of these stakeholders has a critical role to play in achieving the goal of improving representation in clinical research, but the complex nature of the research ecosystem and research processes, combined with lack of accountability and historic underinvestment, means that an issue that should be everyone's responsibility can become no one's priority. In this report, the committee emphasizes that the research supports taking a systematic approach to addressing this issue, one in which all stakeholders take responsibility for the important role they can play in ensuring representation in clinical research participation.

The committee was asked, "Who bears the cost of more inclusive science?" The responsibility (and therefore the cost) will be borne to some extent by all stakeholders in the larger research ecosystem, acting in concert to achieve this larger societal and scientific goal. Those that profit from scientific discovery bear particular responsibility in shouldering the cost of inclusivity. The federal government has a notably prominent role and responsibility in achieving the goal of more inclusive research, as a primary funder of the research enterprise with taxpayer dollars, regulator of the processes of scientific research, gatekeeper to approvals for monetizing scientific discovery, and purchaser of new drugs and devices. More coherence of federal policy to align investment and accountability to achieve the goals of inclusive science is warranted.

In answering the question of who bears the cost of more inclusive science, we must also ask, "Who bears the cost of the current lack of inclusivity?" That cost is large (as evidenced by the analysis in Chapter 2) and is borne disproportionately by underrepresented and historically excluded communities, but saps the health and economic strength of the entire society.

5. **Creating a more equitable future entails a paradigm shift.**

The committee sees the need for both pragmatic approaches and an aspirational vision. To realize a more equitable future, the report epilogue challenges the field to embrace a paradigm shift that moves the balance of power from institutions and puts at the center the priorities, interests, and voices of the community. An ideal clinical trial and clinical research enterprise pursues justice in the science of inclusion through scalable frameworks; expects transparency and accountability; invests more in people, institutions, and communities to drive equity; and invests in the science of community engagement and empowerment. These ideals should be the foundation of the actions that stakeholders take to make sustainable change.

RECOMMENDATIONS

The committee's recommendations focus on tangible actions that must urgently be taken within the context of the existing structures of the clinical research ecosystem in order to achieve the goals of representation and inclusion. Although individual researchers can take many actions to improve health equity in clinical trials and clinical research, as described in Chapter 5, the committee focused on system-level recommendations to drive change on a broader scale. The committee presents 17 recommendations (see Chapter 6) to improve the representation of underrepresented and excluded populations in clinical trials and clinical research and create lasting change.

The urgency of addressing the equity in research participation and the lack of substantial progress despite stated commitments led the committee to propose bold recommendations with potentially far-reaching implications. The committee is aware that the complexity of the U.S. health-care system poses significant challenges to transforming the clinical research system, and these systematic challenges will also influence the implementation of the committee's recommendations. While providing a complete policy assessment for each recommendation was outside of the committee's scope and charge, the committee does not deny that there will be costs—both fiscal and political—associated with the implementation of the recommendations. These costs must be carefully weighed against the potential for long-term benefit. Changing our nation's approach to clinical research may require significant upfront costs to more equitably recruit and retain a diverse group of participants and to hold investigators accountable when they do not meet these goals. In addition, it will require incentivizing sponsors of clinical research to change the status quo. However, based on the committee's expert opinion and the available evidence, the committee believes that implementation of its recommendations is necessary to truly drive significant and sustained change to the clinical research system.

Reporting and Accountability

1. **The Department of Health and Human Services (HHS) should establish an intradepartmental task force on research equity charged with coordinating data collection and developing better accrual tracking systems across federal agencies, including the Food and Drug Administration (FDA), National Institutes of Health (NIH), Centers for Disease Control and Prevention (CDC), Agency for Healthcare Research and Quality (AHRQ), Health Resources Services Administration (HRSA), Indian Health Services (IHS), Centers for Medicare & Medicaid Services (CMS), and two departments outside of HHS, the Department of Veterans Affairs and Department of Defense. This task force should be charged with the following:**
    a. Producing an annual report to Congress on the status of clinical trial and clinical research enrollment in the United States, including the number of patients recruited into clinical studies by phase and condition; their age, sex, gender, race, ethnicity, and trial location (i.e., where participants are recruited); their representativeness of the conditions under investigation; and the research sponsors.
    b. Making data more accessible and transparent throughout the year, such as through a data dashboard that is updated in real time.
    c. Determining what "representativeness" means for protocols and product development plans.

d. Developing explicit guidance on equitable compensation to research participants and their caregivers, including differential compensation for those who will bear a financial burden to participate.
2. The FDA should require study sponsors to submit a detailed recruitment plan no later than at the time of Investigational New Drug and Investigational Device Exemption application submission that explains how they will ensure that the trial population appropriately reflects the demographics of the disease or condition under study and that provides a justification if these enrollment targets do not match the demographics of the intended patient population in the United States.
3. The NIH should standardize the submission of demographic characteristics for trials to ClinicalTrials.gov beyond existing guidelines so that trial characteristics are labeled uniformly across the database and can be easily disaggregated, exported, and analyzed by the public. The data reported should include the number of patients; their age, sex, gender, race, ethnicity, and trial location (i.e., where participants are recruited); who sponsors them; and language accessibility.
4. In grant proposal review, the NIH should formally incorporate considerations of participant representativeness in the score-driving criteria that assess the scientific integrity and overall impact of a grant proposal. These criteria should be part of the assessment of the scientific approach, including whether it is appropriate for generating insights for the populations to whom the results are intended to generalize. The criteria should also be incorporated in the assessment of whether investigative teams and environment have detailed and feasible plans to meet the goals of representative study enrollment. Additionally, the NIH should assess in its annual review of progress reports of funded studies whether a given study has met the proposed enrollment goals of representativeness by race/ethnicity, sex, and gender, and should establish a plan for remediation for the investigator and/or organization that includes criteria for putting funding on hold that has not met predefined recruitment goals.
5. Journal editors, publishers, and the International Committee on Medical Journal Editors should require information on the representativeness of trials and studies for submissions to their journals, particularly relative to the affected population; should consider this information in accepting submissions; and should publish this information for accepted manuscripts. The information required should include the following:
   a. The disease, problem, or condition under investigation.
   b. Special considerations related to sex and gender, age, race or ethnic group, and geography.

c. The overall representativeness of the trial, including how well the study population aligns with the target population in which the results are intended to generalize. If the study population does not align with the population affected by the disease, authors should provide scientific justification for why this is the case.
6. The Office of Human Research Protections (OHRP) and the FDA should direct local institutional review boards (IRBs) to assess and report the representativeness of clinical trials as one measure of sound research design that it requires for the protection of human subjects. Representativeness should be measured by comparing planned trial enrollment to disease prevalence by sex, age, race, ethnicity and trial location (i.e., where participants are recruited). Protocols in which the planned enrollment diverges substantially from disease prevalence should require justification. The OHRP and FDA should establish a plan for remediation for local IRBs that frequently approve protocols that are not representative.
7. The CMS should amend its guidance for coverage with evidence development (CED) to require that study protocols include the following:
   a. A plan for recruiting and retaining participants who are representative of the affected beneficiary population in age, race, ethnicity, sex, and gender
   b. A plan for monitoring achievement of representativeness as described above, and a process for remediation if CED studies are not meeting goals for representativeness

Federal Incentives

8. In order to determine how to take action on the most effective accountability and incentive structures, Congress should direct the FDA to enforce existing accountability measures, as well as establish a taskforce to study new incentives for new drug and device applications for trials that achieve representative enrollment. Incentive programs should be designed to improve representativeness in clinical research, improve clinical outcomes, and ensure they do not reduce access to new therapies. Some ideas include:
   a. Tax incentives, such as tax credits for research and development.
   b. Fast-Track criteria and exemption from some FDA drug application fees.
   c. Extended market exclusivity to sponsors who meet predefined criteria of representativeness.
   d. Refusing to file an application that does not appropriately represent the target population under study.

9. The CMS should expedite coverage decisions for drugs and devices that have been approved based on clinical development programs that are representative of the populations most affected by the treatable condition.
10. The CMS should incentivize community providers to enroll and retain participants in clinical trials by reimbursing for the time and infrastructure that is required. Through the creation of new payment codes, the CMS should reimburse activities associated with clinical trial participation, including but not limited to data collection and personnel (e.g., community health workers, patient navigators) to support research education and recruitment.
11. The Government Accountability Office (GAO) should assess the impact of reimbursing routine care costs associated with clinical trial participation for both Medicare (enacted in 2000) and Medicaid (enacted in 2020). The assessment should include an analysis of whether there is timely and complete reimbursement, any implications for innovation and care delivery to underrepresented populations, and any challenges to implementation.

Remuneration

12. Federal regulatory agencies, including the OHRP, NIH, and FDA, should develop explicit guidance to direct local IRBs on equitable compensation to research participants and their caregivers. In recognition that research participation may pose greater hardship or burdens for historically underrepresented groups, the new guidance should encourage and allow for differential compensation to research participants and their caregivers according to the time and financial burdens of their participation. Differential compensation may include additional reimbursement for expenses including but not limited to lost wages for those with lower socioeconomic status (SES), transportation costs, per diem, dependent care, and housing/lodging where applicable.
13. All sponsors of clinical trials and clinical research (e.g., federal, foundation, private and/or industry) should ensure that trials provide adequate compensation for research participants. This compensation may include additional reimbursement for expenses including but not limited to lost wages for lower SES participants and family caregivers, transportation costs, per diem, dependent care, and housing/lodging where applicable.

Education, Workforce, and Partnerships

14. All entities involved in the conduct of clinical trials and clinical research (academic centers, health-care systems, sponsors, regulatory

agencies, and industry) should ensure a diverse and inclusive workforce, especially in leadership positions.
15. Leaders and faculty of academic medical centers and large health systems should recognize research and professional efforts to advance community-engaged scholarship and other research to enhance the representativeness of clinical trials as areas of excellence for promotion or tenure.
16. Leaders of academic medical centers and large health systems should provide training in community engagement and in principles of diversity, equity, and inclusion for all study investigators, research grants administration, and IRB staff as a part of the required training for any persons engaging in research involving human subjects. This training should incorporate strategies to enhance diverse recruitment and retention in clinical research, as well as planning of and budgeting for these efforts and timely reimbursement of partnering agencies and organizations.
17. HHS should substantially invest in community research infrastructure that will improve representation in clinical trials and clinical research. This funding should go to agencies such as the HRSA, NIH, AHRQ, CDC, and IHS to expand the capacity of community health centers and safety-net hospitals to participate in and initiate clinical research focused on conditions that disproportionately affect the patient populations they serve.

1

The Committee's Task

Throughout history, biomedical research has contributed enormously to progress in treating and preventing disease and overall life expectancy. Such progress requires clinical, translational, and population studies—including clinical trials, observational studies, and implementation designs—in which people volunteer as participants to help researchers find answers to specific questions about how health, disease, and therapeutic interventions work.[1] Studies such as these are critical for ensuring that fundamental discovery translates into improvements in human health. The data and evidence these studies generate also are critical for securing reimbursement—and therefore patient access—to therapeutic interventions. Yet, advancing the nation's capacity to protect and improve health is unobtainable if large swaths of the U.S. population, often those with the greatest health challenges (or most premature morbidity and mortality), are less able to benefit from these discoveries because they are not adequately represented in clinical research studies.

The scientific necessity to improve research equity is urgent. The 2020 U.S. Census found that the number of people who identify as white has decreased for the first time since a census started being taken in 1790 (Bahrampour and Mellnik, 2021), and despite the country becoming more diverse, the nation's health disparities persist. Without major advancements in the inclusion of traditionally underrepresented groups in health research, meaningful reductions in racial and ethnic inequities in chronic diseases such as diabetes, cancer, and Alzheimer's disease remain unlikely. The critical need for research findings to generalize to

---

[1] In many implementation designs there is a waiver of consent, so participants do not overtly volunteer.

the entire U.S.US population has also long been recognized and underscored by the National Institutes of Health (NIH) mandate in the Public Health Service Act, Section 492B (42 U.S.C. § 289a-2), to ensure inclusion of women and minority populations in all NIH-funded clinical research in a manner that is appropriate to the study question; however, more needs to be done to achieve the goals of representation and inclusion.

As described in detail in Chapter 2, lack of representation has a range of serious consequences. Failing to adequately represent underrepresented and excluded populations in clinical trials and research may limit scientific innovation and exacerbate health disparities. Lack of representation can shift the direction of research toward majority groups, feed mistrust, and ultimately may impugn the integrity of the scientific enterprise or of a specific therapeutic or discovery. Gaps in representation are particularly problematic when one considers the stark and deep disparities in disease burden experienced by the same populations that have not been represented in clinical research. In short, lack of representation compromises generalizability of clinical research findings to the U.S.population, hinders innovation, compounds low accrual rates causing trials to fail, leads to lack of effective medical interventions, undermines trust, compounds health disparities, and costs the United States billions of dollars each year.

Some may note that even if clinical trials look like the population affected by the disease under study, the trials likely are not powered to examine subgroup differences. However, inclusion of underrepresented and excluded populations in clinical trials is crucial—even and perhaps especially when these groups are small in number and even in studies that are not adequately powered to draw conclusions about specific populations. The need for equity demands it. In many settings, clinical research represents the best available health-care option. Individuals from underrepresented populations must have the opportunity to access it.

Additionally, studies that are not powered to draw conclusions about subgroups can and do lead to testable hypotheses that can and should be followed up in subsequent research that uses oversampling to achieve adequate power. Studies that do not include individuals from underrepresented populations yield neither knowledge nor testable hypotheses about these groups, and instead perpetuate and exacerbate gaps in access, gaps in knowledge, and disparities in health outcomes.

Although the committee routinely calls for clinical trials to represent the populations affected by the disease under study, there are certainly cases where this would not apply. For example, some trials may oversample certain populations for which we have limited information or where diversity may be limited by the geography of the trial site. The committee's recommendations reflect this nuance, with the opportunity to provide justification when trial enrollment does not match the demographics of the intended patient population. However, the committee believes that in aggregate, clinical trials should look like the populations affected by the disease under study.

While Chapter 2 presents the committee's detailed analysis of these costs and consequences of a lack of inclusion in clinical research, and provides an overview of the current state of representation within the clinical research ecosystem, this chapter describes the committee's process in carrying out the study and the statement of task that guided the committee's work, and it outlines the structure of the report.

COMMITTEE TASK AND APPROACH

In the 2020 appropriations process, Congress mandated that the NIH fund a National Academies of Sciences, Engineering, and Medicine study "examining and quantifying the long-term medical and economic impacts of the inclusion of women and racial and ethnic minority population groups in biomedical research and subsequent translational work" (see Box 1-1). In accordance with this mandate, the National Academies appointed and tasked this diverse committee of experts with carrying out a study that would identify policies, procedures, programs, or projects aimed at increasing the inclusion of these groups in clinical research and the specific strategies used by those conducting clinical trials and clinical and translational research to improve diversity and inclusion. The committee was tasked with modeling the potential economic benefits of full inclusion of men, women, and racial and ethnic groups in clinical research and to highlight new programs and interventions in medical centers and other clinical settings designed to increase participation. The full Statement of Task can be found in Box 1-2.

The Committee on Improving the Representation of Women and Underrepresented Minorities in Clinical Trials and Research consisted of 13 members with a broad range of expertise, including health disparities, health-care policy, health economics, community-engaged research, running diverse clinical trials in academia and biopharma, nursing science, clinical outreach to medically underserved populations, and patient-centered care.

The committee deliberated over five virtual meetings and one hybrid meeting between January 2021 and December 2021, as well as over many conference calls between January 2022 and April 2022. In addition to the closed meetings, the committee held three virtual public workshops, where outside speakers were invited to inform the committee's deliberations and members of the public were invited to comment and ask questions. The speakers provided valuable input on a wide range of topics that helped to inform and shape the committee's approach to the report. In addition, the committee commissioned five papers to address various aspects of the Statement of Task. First, to accomplish the economic analysis, the committee worked with authors to commission an economic analysis using the Future Elderly Model to estimate the social costs of health disparities for groups that have historically been underrepresented in clinical trials and clinical research (see Chapter 2 for a summary of this work and Appendix A for the full analysis). Second, the committee commissioned a paper to estimate the current demographic

> **BOX 1-1**
> **Congressional Mandate**
>
> Language found in the explanatory statement for Division A (Departments of Labor, Health and Human Services, and Education) to the Consolidated Appropriations Act, 2020.
>
> Inclusion in Clinical Research. - The agreement directs NIH [National Institutes of Health] to fund a NASEM [National Academies of Sciences, Engineering, and Medicine] study examining and quantifying the long-term medical and economic impacts of the inclusion of women and racial and ethnic minorities in biomedical research and subsequent translational work, and has provided $1,200,000 to fund this effort. The NIH is directed to report to the Committees on this issue and it should include a review of the existing research on the long-term economic benefits of increasing the participation of women and racial and ethnic minorities in clinical trials and biomedical research, including an analysis of fiscal implications of inclusion on the nation's overall healthcare costs; examine new programs and interventions in medical centers that are currently working to increase participation of women of lower socioeconomic status and women who are members of racial and ethnic minority groups; identify programs that are positively addressing issues of underrepresentation; and analyze whether and how those programs are replicable and scalable; and identify more inclusive institutional and informational policies and procedures to improve health outcomes for racial and ethnic minorities, including health referral forms, continuing education classes, and more.

status of clinical trial and clinical research participants, as this analysis is not readily available in the literature (see Chapter 2 for a summary of this work and Appendix B for the full analysis). Third, the committee commissioned a paper on successful facilitators for having representative clinical trials, based on 20 qualitative interviews conducted in 2021 with research teams (investigators and staff) involved in clinical trials that successfully achieved diverse enrollment (see the full analysis in Appendix C). The remaining two commissioned papers were reviews of the literature: one focusing on federal policies that have influenced the diversity of clinical trials and clinical research (see Chapter 3) and the second focusing on the structural barriers to having representative trial enrollment (forming the basis of Chapter 4). The authors of these papers worked closely with committee members to help scope the papers, develop methodology, and ultimately ensure that the papers informed the report. These commissioned analyses were critical for helping the committee meet its task and form the basis of this report.

The committee notes that although the Statement of Task is rather broad, the task of this study was to focus specifically on women and racial and ethnic minority population groups in clinical trials and clinical research. Although it was out of scope for the committee to examine other excluded and underrepresented populations in clinical trials and clinical research, such as children, older adults,

> **BOX 1-2**
> **Statement of Task**
>
> An ad hoc committee under the auspices of the Committee on Women in Science, Engineering, and Medicine will undertake a study examining the long-term medical and economic impacts of the lack of inclusion of women and underrepresented minority groups in clinical research and subsequent translational work. The study will:
>
> - Review the existing research on the long-term health and economic benefits of increasing the participation of women and racial and ethnic minorities in clinical trials and research, including existing research on the fiscal implications of inclusion on the nation's overall health care costs.
> - Review the existing literature on the factors that affect inclusion, including building equity into research designs and methods, unique inclusion-related challenges of specific medical or behavioral health conditions, and community-driven approaches to research including women and other underrepresented groups.
> - Examine new programs and experimental initiatives in medical centers that are currently working to increase participation of women and members of racial and ethnic minority groups.
> - Highlight programs that are positively addressing issues of underrepresentation in clinical trials, including models to address trust from a patient perspective, and analyze whether and how those programs are replicable and scalable.
> - Identify more inclusive institutional and informational policies and procedures to increase the likelihood of improved health outcomes for women and racial and ethnic minorities, including health referral forms, continuing education classes for practitioners, and more.
>
> The committee will produce a final consensus report.

persons with disabilities, rural and frontier populations, and more, the committee would like to underscore that these are critical topics for future work. In addition, the committee's scope was narrowed to clinical trials and clinical research, but the topic of database-based studies is another important area for future work, as these studies also suffer from representation issues and data-reporting issues.

The committee recognizes the importance of including the participant, patient, and caregiver voice throughout the research process, from study design and recruitment to the dissemination of findings. Centering these voices are essential to conducting research that is more equitable and responsive to the needs of communities. However, the committee faced several structural barriers that limited the representation of these voices in the report, many of which the committee addresses throughout the following chapters, including challenges with remuneration and agency policies. While we recognize the need for a more intentional approach and room for improvement, the committee made efforts to include the perspective of patient advocacy stakeholders throughout the development of this report, including as speakers in the committee's public forums and by including representatives from the patient advocacy community on the committee. In addition, the committee cites studies throughout the report that do include the voices of community members to bring these perspectives from the literature into this report.

Throughout the report, the committee uses examples from specific disease areas to illustrate points and provide evidence. Although these diseases are mentioned throughout the report, it is important to note this study focuses on diversifying clinical trials and clinical research across all disease areas.

DEFINITION OF TERMS

Throughout this report the committee uses several terms that would benefit from definition. *Clinical research* or *medical research* includes (1) research conducted with human subjects or on material of human origin for which direct human participant interaction is needed, including clinical trials; (2) epidemiologic and behavioral studies; and (3) outcomes, health services, and large database research.

The very task of this committee—to examine the long-term medical and economic impacts of including women and racial and ethnic minority populations in clinical trials and biomedical research, and to identify and describe policies and programs that support inclusion—raised and continues to raise critical questions about the meaning of its terms. The committee acknowledges and believes race is a social construct that has, from its inception to the present day, exclusively benefited people who identify as or appear to be white or European and profoundly harmed those who do not. It further acknowledges that the social construction of race, and its racist derivatives have created measurable, sustained, and life-threatening biological outcomes. To meet our statement of task, and to look beyond it toward a just and equitable society, we use the imperfect language we have to describe and offer resolutions for observable inequities in health outcomes and clinical research. To that end, the committee chose the term *underrepresented and excluded populations* as the broadest term to refer to the populations and communities that are the focus of this report. The term "Underrepresented" calls attention to studies, research foci, funding streams, and other components of the research ecosystem that draw from populations whose demographic characteristics are not representative of the people who ought to benefit from it. "Excluded" emphasizes that the choices and actions of various entities in the research ecosystem result in de facto exclusion of people from underrepresented groups, even when these individuals meet inclusion criteria. We also note that evidence-based exclusion criteria (such as for pregnant and lactating individuals) can result in underrepresentation.[2] So while most women

---

[2] The committee recognizes that exclusion criteria are necessary, and that the inclusion of pregnant and lactating individuals may require special consideration, as discussed at length in Chapter 2.

and racial and ethnic minority population groups are technically included, they are often underenrolled. Thus, in this report the committee shifted the focus away from inclusion toward representativeness, which is defined as matching the self-reported demographics of those enrolled in clinical trials and clinical research to the demographic characteristics of the population affected by the particular illness or condition under study, including self-reported age, sex, gender, race, ethnicity, and socioeconomic status.

To describe excluded and underrepresented populations, the committee uses the terms *Black, white, Latinx, American Indian/Alaska Native, Native Hawaiian/Pacific Islander*, and *Asian American* in the broadest sense, acknowledging that these terms do not capture the complexity and intersectionality between and within these groups. When possible, more specific terms are used. However, when describing published research, the committee uses the same language as the referenced publication to ensure accuracy. Therefore, the reader may see inconsistencies in terminology throughout the report due to inconsistencies in language throughout the literature. Literature on other underrepresented groups, including the elderly, LGBTQIA+ individuals, transgender/gender nonbinary individuals, and residents of rural and frontier areas, is not nearly as complete or detailed as that for ethnic/racial groups. The literature that does exist is most often at the intersection of these groups with the ethnic/racial minority groups of which they are a part (older Hmong populations, Black women). The committee therefore discusses these groups only when relevant and/or specific data are available.

ORGANIZATION OF THE REPORT

In the chapters that follow, the committee provides an overview of the threats posed by lack of representation in clinical studies, the current status of representation in clinical research participation, and the clinical research ecosystem (Chapter 2); offers an overview of the existing landscape of current and past federal policies and practices aimed at addressing this issue (Chapter 3); outlines the range of barriers to full inclusion of underrepresented and excluded populations in clinical trials and research (Chapter 4); identifies facilitators of successful inclusion in clinical research (Chapter 5); and summarizes key principles and recommendations for how a range of stakeholders can take action to address this critical national issue (Chapter 6).

The committee's work focuses on tangible actions that must urgently be taken within the context of the existing structures of the clinical research ecosystem in order to achieve the goals of representation and inclusion. In addition, the committee recognizes that a more transformative and equitable future is possible and desirable; the epilogue describes such a potential vision.

2

Why Diverse Representation in Clinical Research Matters and the Current State of Representation within the Clinical Research Ecosystem

The analysis draws substantially from research papers by Dr. Bryan Tysinger, Ph.D. and Jakub P. Hlávka, Ph.D. which were commissioned for this study. The full research papers can be found in Appendix A and Appendix B.

In this chapter, the first section details how lack of representation risks undermining the overall goals of clinical research and the costs of maintaining the status quo. The next section describes the current status of clinical research representation with a focus on women and racial/ethnic minority populations. The chapter ends with a description of the clinical research ecosystem with a focus on the processes that might address diverse representativeness.

LACK OF REPRESENTATION IN CLINICAL RESEARCH THREATENS THE OVERARCHING GOALS OF CLINICAL RESEARCH

While inclusion of women and historically excluded groups in clinical research has long been viewed as a worthy aim, what are the consequences of failure to achieve this aim? As the overarching goal of the U.S. investment in biomedical research is to improve the health and well-being of the entire U.S. population, the committee identified seven potential threats to this goal posed by lack of representation in clinical research.

1. **Lack of representation compromises generalizability of clinical research findings to the U.S. population.** Over the latter half of the 20th century, randomized controlled trials (RCTs) came to be regarded by the

medical community as the gold standard in evidence-based medicine to determine the safety and efficacy of investigational medical therapies. Initially, the results from these RCTs were largely considered to be generalizable to all patient populations (Bothwell et al., 2016). Over the past few decades, growing evidence has surfaced to challenge that assumption (Sirugo et al., 2019). Specifically, research has demonstrated that many groups underrepresented and excluded in clinical research can have distinct disease presentations or health circumstances that affect how they will respond to an investigational drug or therapy (Beglinger, 2008; Crawley et al., 2003; Garcia et al., 2016; Ramamoorthy et al., 2015). Such differences contribute to variable therapeutic responses and necessitate targeted efficacy and safety evaluation (see Box 2-1). For instance, it appears that men are more likely to respond to tricyclic antidepressants and women to selective serotonin reuptake inhibitors as treatment for depression (Baca et al., 2004; Bano et al., 2004; Kornstein et al., 2000). Reduced renal and hepatic clearance in older adults increases the risk of harms from drugs such as anticoagulants and psychotropic agents (Maixner et al., 1999; Shepherd et al., 1977; Soejima et al., 2022).

Representation by self-identified race and ethnicity is important to generalizability of study findings, but interpretation requires clarity of thought. Racial categories are socially constructed and do not have a biological basis, as is noted at the outset of this report. Some genetic factors that may result in heterogeneity in drug response may be more common in certain ancestral populations, which may be associated with self-identified race and ethnicity, as is the case for classes of medications that have narrow therapeutic window such as anti-coagulants (Box 2-1) or efavirenz (Cummins et al., 2015; Torgensen et al., 2019). In these cases, studies with participants diverse by self-identified race may allow for the identification of specific genotypes important for understanding heterogeneity in drug response. Self-identified race and ethnicity may also be associated with lived experiences that themselves result in specific biological manifestations that are not genetic in origin. For example, the lived experience of structural and interpersonal racism, lower socioeconomic status, and lower educational attainment all appear to be associated with elevations in blood pressure and cardiovascular risk (Hamad et al., 2019, 2020; King et al., 2021; Krieger and Sidney, 2011). Non-genetic factors may affect each population differently and also are subject to epigenetic effects that may vary across populations. Therefore, these analyses are complex and demand nuanced analyses with detailed and high-quality measures on genetic and non-genetic factors, and interpretation of population-specific data in clinical trials. Ensuring diverse participation in scientific studies allow for exploration of all of these factors and their interactions and is critical to the interpretations that allow for generalizability of findings to the population.

> **BOX 2-1**
> **Example: Adjustment of Dosing for Warfarin**
>
> The therapeutic experience with warfarin offers a cautionary tale. When clots form in blood vessels, they can detach and obstruct blood flow—sometimes resulting in strokes or pulmonary embolism. Warfarin can prevent these deadly thromboembolisms by inhibiting clotting within the blood.
> However, too much warfarin can have an adverse side effect—namely, excess bleeding—and there is a 20-fold interpatient variability in therapeutic warfarin dose requirements. As a result, warfarin is one of the leading causes of adverse drug events, and incorrect dosing can lead to increased risk of bleeding, hospitalization, and death.
> Nearly half of the variability in patient response can be explained by genetic variants—both in the drug metabolizing enzyme (CYP2C9) and the drug target (VKORC1) of warfarin. The therapeutic requirements for warfarin differ by the presence of these specific genetic variants, and their frequencies vary substantially across genetic ancestries. For example, populations with greater genetic African ancestry are more likely to require higher average daily doses of warfarin (about 6 mg per day), whereas populations with greater genetic Asian ancestry require lower average warfarin doses (about 3.4 mg per day).
> However, because most of the early genetic studies of warfarin were conducted in populations with predominantly European ancestry, dosing algorithms failed to adequately generalize to the diverse U.S. population. Indeed, even though warfarin has been approved for human use since 1951, it was not until 2013 that it was learned that genotype-guided dosing would be of clinical utility.
>
> SOURCE: Drozda et al., 2015.

Threats to generalizability exist for all studies, not just clinical trials of new therapeutics. For example, implementation of evidence-based practice in community settings may be limited because the practice sites may be substantially different from those included in clinical research studies. Clinical research is often performed in well-resourced tertiary care sites in large urban centers, and may have limited applicability to community sites, less well-resourced safety net settings, and rural settings. Genetic and genomic studies that form the basis for "precision medicine" are increasingly recognized to be built on data from mostly populations of European descent (Martin et al., 2019; Sirugo et al., 2019). Tools such as polygenic risk scores that may help to identify risk and target therapeutic agents more selectively are recognized to be substantially less effective in populations with different genetic ancestry (Martin et al., 2019).

2. **Lack of representation costs hundreds of billions of dollars.** It is important to also quantify the potential economic benefits of greater inclusion in clinical trials. The committee commissioned a study using

the Future Elderly Model (FEM), a model developed over more than two decades with funding from the National Institutes of Health, MacArthur Foundation, Centers for Medicare & Medicaid Services (CMS), and Department of Labor (see Boxes 2-2 and 2-3).

For the committee's report, the model follows a representative cohort of Americans over time, generating snapshots of their health, functional status, and medical spending. Health is measured based on a set of self-reported chronic diseases such as diabetes, hypertension, heart disease, cancer, and other conditions. At any point in time, a person's health and functional status is translated into a disability-adjusted life-year and assigned a monetary value ($150,000 per disability-adjusted life-year). In this way, the model captures both how long a person will live, and the years of disability-free life they will experience over their lifetime. (See Appendix A for more detail.)

Using this model, the committee estimated the social costs of health disparities for groups that have historically been underrepresented in clinical trials and in clinical research. The presumption is that disparities in three outcomes could potentially have been mitigated if clinical trials had been more inclusive: quantity of life (measured by life expectancy), quality of life (measured by disability-free life), and working life (measured by years in the labor force) (see Box 2-1). To quantify the potential benefits of more diverse representation, the committee identified six historically underrepresented groups with sufficient sample size to support the analysis. Throughout, non-Hispanic white men served as the reference group due to their historical inclusion and representation in clinical trials. Self-reported non-Hispanic Black men, Hispanic/Latinx men, non-Hispanic white women, non-Hispanic Black women, and Hispanic/Latinx women all potentially benefit from narrowing the differential impact of disease on the outcomes of interest (see Table 2-1).

The committee then considered potential benefits of reducing disparities in three key chronic diseases: diabetes, heart disease, and hypertension. In aggregate, the committee found when using the FEM that health disparities in diabetes will cost society more than $5 trillion through 2050—including mortality, morbidity, and loss of work. Heart disease would cost more than $6 trillion, and hypertension even more.

What accounts for these differences? Much of it has to do with the shorter life expectancy for Black and Latinx populations with these diseases. The United States has seen dramatic changes in population health over the last century—driving an increase in life expectancy and productivity. As a result, many people have enjoyed greater overall wealth, much of which was previously sapped by illness, disability, and premature death. However, these gains have been uneven (Jamison et al., 2013).

BOX 2-2
Economic Cost of Lack of Representation in Clinical Trials and Research

Lack of equal representation in clinical trials has consequences on health outcomes and may contribute to persistent health disparities in the United States. The committee utilized the Future Elderly Model (FEM) to value how chronic conditions differentially affect the lives of older Americans. The FEM is an economic-demographic dynamic microsimulation developed with support from many federal sponsors—including the National Institute on Aging, the Centers for Medicare & Medicaid Services, and the Department of Labor. This model has been used previously by the National Academy of Sciences.

The committee looked at five underrepresented groups (self-reported non-Hispanic Black females and males, Hispanic females and males, and non-Hispanic females) in the U.S. population that will be above age 50 between 2020 and 2050. This represents more than 150 million people. The committee estimates the additional life expectancy, disability-free life expectancy, and working years underrepresented groups could gain from eliminating the disparities relative to self-reported non-Hispanic white males in outcomes from diabetes, heart disease, and hypertension.

- Eliminating all life expectancy disparities for these three common conditions has a total value to society of approximately $11 trillion.[a]
- Eliminating diabetes disparities increases underrepresented groups' life expectancy by almost 1 year for underrepresented groups (an average of 0.87 life years across groups). Disability-free life years also would improve by more than 1 year (an average of 1.09 disability-free life years), and they would remain in the workforce longer (an average of 0.49 years).
- Eliminating heart disease disparities increases more than 1 year for the underrepresented groups (1.04 years on average). Disability-free life years increase nearly 1.5 years (1.49 years on average). Years working increase about a third of a year (0.34 years).
- Eliminating hypertension increases life expectancy nearly 1 year when it is eliminated in the underrepresented groups (0.95 years on average). Disability-free life years increase about 1.5 years (1.51 disability-free life years on average). Years working increase about three-tenths of a year (0.31 years on average).
- Given the assumption that better representation in clinical trials would be able to eliminate even modest reductions in health disparities, the value to society of better representation in these three conditions would be worth billions of dollars.

We estimated the additional impact these chronic diseases have on longevity, years without a disability, and workforce participation for these groups using the Health and Retirement Study, then applied those estimates using the FEM. We valued each additional year of life expectancy or disability-free life expectancy at $150,000 per year, discounted at 3 percent.

The full, detailed analysis can be found in Appendix A.

---

[a] Key calculations include the lifetime risk of diabetes, heart disease, and hypertension, based on projections using the FEM for the underrepresented groups.
SOURCES: Goldman and Orszag, 2014; NASEM, 2015.

> **Box 2-3**
> **Future Elderly Model**
>
> The Future Elderly Model (FEM) and its successor models simulate health and economic consequences for individuals over a lifetime. As such, they have the crucial advantage that they can predict outcomes for certain demographic subgroups of interest to the committee. The FEM has undergone extensive validation. A recent study found it performs at least as well as actuarial forecasts of mortality, while providing policy simulation features not available in actuarial models—including estimates of quality of life.
>
> The FEM and its international counterparts have been used to assess population health disparities in the United States, Asia, and Europe. Indeed, the National Academies relied extensively on the FEM in its report on disparities to predict how long people live and the implications for federal policy. Globally, the Organization for Economic Co-operation and Development relied on these methods in its global report on unequal aging.
>
> SOURCES: Ermini Leaf et al., 2021; NASEM, 2015.

Understanding these consequences requires a broader measure of welfare (Clark, 2013). Returns on social investments are usually measured by economic capacity—particularly gross domestic product in national income accounts. However, such accounts only measure the extent of market activities in an economy. They do not account for other valuable nonmarket activities, of which health is likely the most important (Becker et al., 2005).

The committee's approach measures how health investments could more broadly contribute to social value. These methods have been used to assess progress internationally as well. It does so by measuring the benefits of reducing disparities—which translates into both longer and healthier lives. These methods, which quantify the health improvements in terms of dollars, allow us to compare compressed inequality to other economic outcomes (Goldman et al., 2009; Goldman et al., 2013; Lowsky et al., 2014; Olshansky et al., 2009; Olshansky et al., 2012).

All told, health disparities incur a substantial economic toll on the U.S. society. Of course, better representation in clinical research will not completely alleviate these disparities—after all, they have many interconnected and interdependent causes. However, to the extent that representation in clinical research may improve generalizability of scientific findings across a range of clinical studies for these important health states, drive new discoveries and increase innovation, improve access, and increase trust, representative clinical research may play a role in alleviating these inequities. Even if only 1 percent of these health disparities could be al-

TABLE 2-1 Adjusted Relative Risks for Key Parameters of Interest with 95% Confidence Intervals

	Diabetes			Heart Disease			Hypertension		
	Mortality	Disability	Work	Mortality	Disability	Work	Mortality	Disability	Work
White males	1.00	1.00	1.00	1.00	1.00	1.00	1.00	1.00	1.00
Black males	1.10 [1.02, 1.18]	1.12 [1.07, 1.16]	0.89 [0.85, 0.92]	1.14 [1.07, 1.22]	1.23 [1.18, 1.27]	0.86 [0.83, 0.90]	1.10 [1.02, 1.19]	1.17 [1.13, 1.22]	0.95 [0.93, 0.98]
Hispanic males	1.11 [1.02, 1.20]	1.12 [1.07, 1.16]	0.91 [0.88, 0.94]	1.15 [1.07, 1.23]	1.22 [1.18, 1.27]	0.89 [0.86, 0.92]	1.11 [1.03, 1.20]	1.17 [1.12, 1.21]	0.96 [0.94, 0.98]
White females	1.10 [1.02, 1.19]	1.11 [1.07, 1.16]	0.89 [0.85, 0.92]	1.14 [1.07, 1.21]	1.21 [1.17, 1.26]	0.86 [0.82, 0.90]	1.10 [1.02, 1.18]	1.16 [1.12, 1.20]	0.95 [0.92, 0.98]
Black females	1.11 [1.02, 1.20]	1.10 [1.06, 1.14]	0.88 [0.85, 0.92]	1.15 [1.07, 1.23]	1.19 [1.15, 1.22]	0.86 [0.83, 0.90]	1.11 [1.03, 1.20]	1.15 [1.11, 1.19]	0.95 [0.93, 0.98]
Hispanic females	1.11 [1.02, 1.21]	1.10 [1.06, 1.14]	0.88 [0.85, 0.92]	1.15 [1.07, 1.23]	1.18 [1.15, 1.22]	0.86 [0.82, 0.90]	1.11 [1.03, 1.20]	1.14 [1.11, 1.18]	0.95 [0.92, 0.98]

NOTE: Adjusted relative risks for the key parameters of interest (the underrepresented group and disease interaction term) are shown here. The reference group, non-Hispanic white males, will always have values of 1.0. Relative to white males, being in an underrepresented group and having diabetes is associated with an increase in mortality of 10 to 11 percent, an increase in disability of 10 to 12 percent, and a decrease in workforce participation of 9 to 12 percent. Heart disease is associated with a mortality increase of 14 to 15 percent, an increase in disability of 19 to 23 percent, and a decrease in workforce participation of 11 to 14 percent. Hypertension is associated with an increase in mortality of 10 to 11 percent, an increase in disability of 14 to 17 percent, and a decrease in workforce participation of 4 to 5 percent.

leviated by better representation in clinical research—like the warfarin example in Box 2-1—the analysis shows it would result in more than $40 billion in gains for diabetes and $60 billion for heart disease alone.

These findings suggest that even modest reductions in health disparities as a result of better representation in clinical trials for diabetes and heart disease would result in billions of dollars of savings to U.S. society. Expanding this estimate by alleviating the health disparity more fully, adding other diseases like Alzheimer's disease or cancer, or computing across future cohorts would only add to the potential benefits of better representation.

3. **Lack of representation may hinder innovation.** Diversity in study participants allows for greater exploration of variation in the overall effectiveness of a particular intervention. Exploring "heterogeneity of treatment effects" may be necessary not only to understand variation that affects safety and effectiveness of an intervention in the populations that have been underrepresented in studies but also to identify new biological processes that may, in turn, lead to new discoveries important for all populations. For example, the discovery of proprotein convertase subtilisin/kexin type 9 (PCSK9) has transformed the understanding of cholesterol homeostasis and led to development of important therapeutics for prevention and treatment of atherosclerotic cardiovascular disease (Warden et al., 2020). PCSK9 was discovered while examining differences in cholesterol metabolism in the Atherosclerotic Risk in Communities (ARIC) Study that was specifically designed to investigate variation in cardiovascular risk factors, medical care, and disease by self-reported race, gender, and location (Cohen et al., 2006; UNC, 2022). Researchers found that 2 percent of Black subjects in the ARIC cohort had one of two mutations in PCSK9 that are associated with a 40 percent reduction in low-density lipoprotein, or LDL, cholesterol. These mutations are rare among white people, and therefore, PCSK9 may not have been a target for exploration had diversity not been present in the ARIC study.

4. **Lack of representation may compound low accrual that causes many trials to fail.** According to an analysis by GlobalData (2021), low accrual was the cause for stopping 55 percent of all Phase I–IV clinical trials that were terminated, suspended, or discontinued during 2008–2017 (and for which a reason was given). Improving participation of underrepresented groups would be one way to increase enrollment. Thus, increasing enrollment of underrepresented populations would help solve the number one problem that causes clinical trials to fail, while also helping to ensure clinical data that are more representative of the whole population that could benefit from a studied intervention.

Moreover, improving representation in a way that increases the overall numbers of people who enroll in studies would reduce inefficiency and waste caused by premature study termination. When a study fails to accrue, we often learn little or nothing about the investigational intervention, yet human and monetary resources have been sunk into designing, launching, and maintaining the study.

5. **Lack of representation may lead to lack of access to effective medical interventions.** Approval and indications for new therapeutics are often restricted to the demographics of the populations included in the clinical studies. Lack of representation may thus impede access to a specific therapeutic agent. For example, when Gilead Sciences Inc. sought Food and Drug Administration (FDA) approval for use of its HIV drug Descovy (emtricitabine/tenofovir alafenamide) as pre-exposure prophylaxis (PrEP), the company included only cisgender men and transgender women in its Phase III PrEP study, and presented the FDA with an extrapolation of data from two Phase I pharmacokinetic studies to support approval of the drug for cisgender women. As a result, the label explicitly excludes from the PrEP indication "individuals at risk of HIV-1 from receptive vaginal sex because effectiveness in this population has not been evaluated" (FDA, n.d.). This exclusion is included in direct-to-consumer advertising for the drug, which notes that Descovy for PrEP is not for use in people assigned female at birth.

Guideline-making bodies that synthesize various lines of evidence are often limited in making evidence-based recommendations that apply to all populations when the evidence base on specific populations does not exist; when these recommendations are tied to insurance coverage, these gaps may affect reimbursement of, and therefore access to, health care. For example, the U.S. Preventive Services Task Force (USPSTF) makes evidence-based recommendations for clinical preventive services, and its top-tier recommendations are linked to first-dollar insurance coverage from commercial payers under the Patient Protection and Affordable Care Act (P.L. 111-148). Lack of representative studies on screening for cancer or cardiometabolic disease may lead to recommendations that fail to consider earlier ages or lower biomarker thresholds to start screening that might be warranted in some populations (e.g., lower BMI [body mass index] or earlier age to start diabetes screening in Asian, Black, or Latinx populations; earlier age to start screening for lung, colorectal, breast, or prostate cancer in some populations). For example, in the 2021 USPSTF Report to Congress, the USPSTF was not "able to make a separate, specific recommendation on colorectal cancer screening in Black adults" because of "limited available empirical evidence" despite Black adults having the highest rates of incidence and mortality from colorectal cancer compared with other racial/ethnic groups (USPSTF, 2021). Although other national guideline organizations have historically recommended that Black adults begin screening at an earlier age, the task force recommends all adults start screening at age 45 due to lack of studies that report findings by race.

Clinical trials are a significant, and sometimes the only point of access for the most cutting-edge therapies for advanced disease (e.g., immunotherapy for cancer treatment). Lack of inclusion in clinical trials for advanced therapeutics may result in lack of access to these life-saving interventions.

6. **Lack of representation may undermine trust.** Distrust of the clinical research enterprise and medical establishment rooted in historical and contemporary abuse has been documented as a barrier to participation in clinical studies among some populations. More contemporary work has focused on the importance of the research and medical enterprise working to regain trust and become more trustworthy partners (Alsan and Eichmeyer, 2021; Lucero et al., 2020; Wilkins, 2018). Efforts to overcome barriers to participation in scientific studies and working to create more representative and inclusive research environments may work to increase trust in science and medicine. Studies of vaccine hesitancy for influenza vaccines in Black populations found that knowledge and trust in the process of vaccine development and testing was associated with a higher degree of vaccine uptake (Quinn et al., 2018). Studies have also shown similar trust issues with the SAR-CoV-2 vaccine. Although COVID-19 vaccine trials were some of the most historically diverse trials, one study found that Black participants did not trust that the vaccine results were generalizable to them, contributing to vaccine hesitancy (Bazan and Akgün, 2021). Further, the lack of inclusion of pregnant people in the clinical trials of the SAR-CoV-2 vaccines led to lack of clarity on the use of these vaccines in pregnant people and may have contributed to vaccine hesitancy, even as subsequent data emerged on the importance of preventing COVID-19 infection during pregnancy (Rubin, 2021).

7. **Lack of representation compounds health disparities in the populations currently underrepresented in clinical trials and clinical research.** Healthy People 2020 defines a *health disparity* as "a particular type of health difference that is closely linked with social, economic, and/or environmental disadvantage. Health disparities adversely affect groups of people who have systematically experienced greater obstacles to health based on their racial or ethnic group; religion; socioeconomic status; gender; age; mental health; cognitive, sensory, or physical disability; sexual orientation or gender identity; geographic location; or other characteristics historically linked to discrimination or exclusion." Health disparities are pervasive and prevent us from achieving *health equity*, defined as the "attainment of the highest level of health for all people. Achieving health equity requires valuing everyone equally with focused and ongoing societal efforts to address avoidable inequalities, historical and contemporary injustices, and the elimination of health and health care disparities" (CDC, 2020).

While achieving health equity and reducing health disparities requires far more than just equitable representation in clinical research, failure to achieve equity on this dimension leaves health disparities unaddressed and reinforces inequities.

For example, prostate cancer is the most common cancer in U.S. men. Disparities in prostate cancer incidence and outcomes are particularly prominent. Black men in the United States have a 1.5 times greater chance of developing prostate cancer and are 2.2 times more likely to die from the disease than white men; roughly 30 percent of all prostate cancer deaths in the United States are in Black men. While the nature of these disparities is complex, the fact that Black men make up less than 3 percent of the participants in clinical trials of this common cancer may directly contribute to disparities via the threats listed above (Borno et al., 2019).

CLINICAL TRIALS PRODUCTION PROCESS

Several stakeholders are involved in the process of diversifying trials. While all of these are outlined in sections throughout this report, the committee thought it was helpful to detail the various stakeholders and processes in one place, for the ease of the reader (see Figure 2-1).

To begin from the ideation stage, investigators, whether in industry or academic medical centers, are often the ones developing research questions. However, working with community organizations and community partners helps build relationships, ensures that the research resources align with local needs, and helps to recruit and retain study participants in the research. The research itself is funded by federal sponsors (the National Institutes of Health [NIH], Centers for Disease Control and Prevention [CDC], Veterans Administration, Department of Defense, and Agency for Healthcare Research and Quality), nongovernmental organizations (academic institutions, patient advocacy groups, and philanthropic organizations), and industry (pharmaceutical, biotechnology, and medical device companies), and the priorities of these funders heavily influence the research that is done.

The study design process is heavily influenced by access to health care and clinical research, as well as the location of where the study itself is completed. Study participants are recruited and selected through word of mouth and social networks, through their primary care physician, and through contract research organizations. Recruitment of participants is highly dependent on where the study itself is taking place and the populations that access care at those sites. For example, federally qualified health centers are more likely to serve uninsured and impoverished families than other sites, such as academic medical centers and private practitioners. The committee would also like to note that while this report largely focuses on clinical trials and clinical research in the United States, many trials are completed overseas. In 2010, the FDA completed a report on the extent to which data submitted to the FDA was from foreign clinical trials. It found that 80 percent of approved applications for drugs and biologics contained data from clinical trials and that over half of clinical trial subjects and sites were located outside of the United States (HHS, 2010).

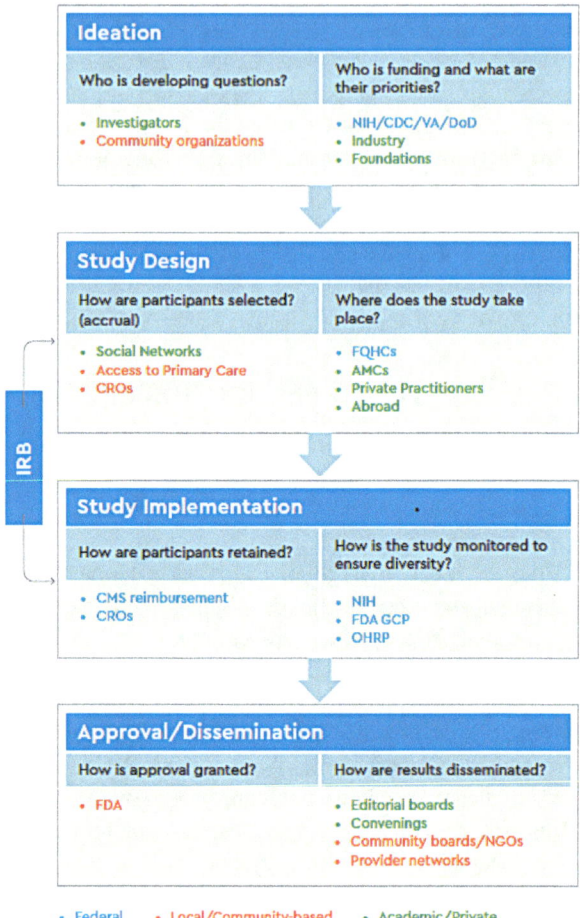

FIGURE 2-1 Overview of the clinical trial ecosystem.

The study implementation phase is heavily influenced by insurance coverage and regulatory bodies. Federal agencies, such as the FDA, have the authority to ensure that trials are diverse and representative. Offices such as the Office for Human Research Protections do not oversee individual studies, but have the authority to revoke assurance within an institution if it is out of compliance with the Common Rule (45 CFR 46). If the NIH funds the study, it also has monitoring authority, along with private sponsors of trials. Participants are often retained assuming the cost of their participation and related costs are reimbursed by the CMS. Institutional review boards (IRBs) are also heavily influential in this stage, although they also influence the study design process as well.

Lastly, the approval and dissemination phase influences the diversity of trials. If investigators would like to submit for drug or biologic approval to the FDA, they have to submit the trial design and trial population. Even if they are not seeking FDA approval, medical journals and editorial boards and scientific conferences ultimately decide what information is published and disseminate results to the public. Ultimately, health-care purchases and payers (e.g., Medicare, private insurers, and individual patients) are primary consumers of clinical trial results, since this informs coverage and health decisions for individuals.

CURRENT STATUS OF CLINICAL TRIAL AND CLINICAL RESEARCH PARTICIPATION: LITTLE CHANGE OVER TIME

In the past three decades, diversity in clinical trials has become an important policy priority advanced by federal agencies such as the NIH Office of Research on Women's Health, the FDA Office of Women's Health and the Society for Women's Health Research, and later by the FDA Office of Minority Health and Health Equity (FDA, 2011a) (see Chapter 3 for extensive analysis of the policy landscape). Despite the increased focus on the lack of women and historically underrepresented populations in U.S.-based clinical trials and research, research participants remain mostly white and male (Coakley et al., 2012; Wissing et al., 2014; Nature Medicine, 2018). Although contemporary reviews have shown increases in participation of women, and more modest increases in participation of racial and ethnic minority population groups and older populations, substantial and significant underrepresentation remains, particularly within certain medical disciplines and diseases, including cardiology, oncology, Alzheimer's Disease, and HIV/AIDS (Chen et al., 2014; Tahhan et al., 2018; Huamani et al., 2019; Ashford et al., 2020; Tahhan et al., 2020; Reihl et al., 2021). Further, even though women's representation in trials has increased, knowledge gaps remain, especially regarding treatment during pregnancy and while lactating (Geller et al., 2011; Scott et al., 2018; Vitale et al., 2017). It is also not clear from available data whether increases in women's representation in clinical trials, writ large, is being driven by clinical study of diseases and conditions that disproportionately affect women.

An FDA summary report of clinical trials of drugs conducted between 2015 and 2019 shows that non-Hispanic white populations compose 78 percent of participants enrolled in U.S. trial sites (FDA, 2020a), though they comprise 61 percent of the country's population (Ortman and Guarneri, 2009). Although it is not the focus of this report, it is important to note that there are additional issues of underrepresentation for age, such as for children and older adults, where issues of informed consent remain a barrier (Committee on Drugs, 2014; Zulman et al., 2011). The continued lack of representation is seen across numerous fields of medical research: different studies have found that racial and ethnic minority population groups and women remain underrepresented in oncology (Chen et al., 2014; Reihl et al., 2021), cardiovascular (Kim et al., 2008), ophthalmology (Berkowitz et al.,

2021), and surgical trials (Kalliainen et al., 2018). Further, when clinical trials do include underrepresented populations, subgroup-specific analyses and results are oftentimes missing or poorly executed (Assmann et al., 2000; Wang et al., 2007).

Transparency and Accountability in Participation

In 2015 the FDA published its first Drug Trial Snapshots, reporting on the demographic characteristics of participants in studies that resulted in product approvals the same year. The Snapshots made clear the extent to which underrepresented and excluded populations were underrepresented in trials for products that may eventually be prescribed or used in their medical care. In response to this and other recent documentation of the homogeneity of clinical trial participants, bioethicists, scientists, and funders have turned their attention to issues of transparency and accountability (Bierer, 2020; Hudson et al., 2016).

The National Institutes of Health aimed to improve transparency in the entire research enterprise by establishing a systematic process for tracking research studies from application through dissemination of results. The process includes the following key elements: requiring Good Clinical Practice training for investigators and staff; requiring investigators to submit clinical trial applications to trial-specific funding opportunity announcements (FOAs), which require more detailed descriptions of trial design, recruitment, and retention, and analyses plans compared with more generic parent FOAs; requiring more specific notices of award, which describe principal investigator responsibilities for publication of results and data sharing, where applicable; establishing a single IRB requirement to prevent delays in study implementation; and requiring clinical trial registration and summary results with financial penalties for failure to comply (Hudson et al., 2016).

To get a more up-to-date picture of who is participating in clinical trials, the committee commissioned research to analyze two different data sets for the trends in participant inclusion by sex and race/ethnicity in clinical trials (see Appendix B for full analysis). The first, the FDA Drug Trial Snapshots data, includes demographic data on trials from all *approved* drugs between 2014 and May 1, 2021 (FDA, 2021b). The second is demographic data on all NIH-funded clinical research and Phase III clinical trials from each institute and center at the NIH, from 2013 to 2018 for which data are available. The results of the literature review, as well as the commissioned analyses, are reviewed below for gender and race/ethnicity.

Although improving representation of women and historically excluded groups has been a priority at the NIH and FDA and other federal agencies, the committee noted that limited systematic reporting on the state of participation in clinical trials and clinical research is accessible in the public domain. For example, although the NIH now reports clinical trial enrollment in NIH-sponsored trials by research, condition, and disease categories (starting in 2018), there are only data available for 2018 at the time of the writing of this report, which did not allow the committee to do a longitudinal assessment of enrollment in clinical trials (NIH, 2022). Additionally, although data sets can be downloaded from ClinicalTrials. gov, the committee faced challenges with the consistency of the data reporting and could not extract demographic data from the database in a reliable fashion. The committee spoke to researchers who have published on demographic data of trials in a specific disease area using ClinicalTrials.gov, and they faced similar issues, requiring researchers to manually insert the data from ClinicalTrials.gov into their own database using a subset of trials (Ludmir et al., 2019). Since the committee was looking at trial participation across all disease areas and over time, manually entering these data would not have been possible. To do the analysis in this report, staff searched through individual NIH institutes' biennial and triennial reports and manually entered the reported data for each from 2013 to 2018. Although this gave an idea of demographic trends over time for individual institutes, there were inconsistencies in the way the data were reported, particularly for reporting ethnicity and gaps for certain years, making the data difficult to analyze and very labor intensive. Further, several analyses have shown discrepancies between self-reported and electronic health record responses to race and ethnicity data particularly for participants who do not identify as white (Azar et al., 2012; Boehmer et al., 2002). Inaccurate reporting of race and ethnicity data impedes the ability to examine health inequities driven by the social construction of race. Therefore, enrolling participants using self-reported race and ethnicity and not guessing based on presentation is a more reliable way of reporting race and ethnicity in these databases, yet it is unknown how race or ethnicity was determined in reporting to ClinicalTrials.gov. Thus, the success of efforts to improve representation in clinical trials and clinical research is difficult to fully evaluate.

Gender Diversity in Clinical Trials

Despite the regulatory efforts to increase gender diversity in trial enrollment, evidence from the 1990s and early 2000s suggested relative underrepresentation of women and racial and ethnic minority population groups in clinical trials (Cotton, 1990; Harris and Douglas, 2000; Mak et al., 2007; Murthy et al., 2004), which persisted until 2016, when women surpassed men in their participation in clinical trials (FDA, 2020a). However, the overwhelming majority of women participating in clinical trials in the United States are white women (78 percent between 2015 and 2019), and trials routinely exclude pregnant and lactating individuals from participating (FDA, 2019).

Status of Women's Participation in Clinical Trials

The slow progress is particularly significant given that sex differences are observed in response to some drugs, including the prevalence of adverse events (Anderson, 2005; FDA, 2011a). Recent work has confirmed the challenge of enrolling women in some therapeutic areas: in stroke clinical trials, for instance, women have been underrepresented even after incidence and prevalence of the disease is taken into account (Carcel and Reeves, 2021), with highest underrepresentation reported in secondary prevention trials (10 percent in one study) (Strong et al., 2020).

Other data do suggest improvements in some areas. For example, somewhat optimistic results were described in a subset of trials studied by Eshera et al. (2015): in studies of drugs approved between 2010 and 2012, just 45 percent of trial participants in small molecule trials were women, but women represented 65 percent of participants in biologic trials. The authors concluded that 82 percent of trials had a study population representative of the sex distribution in the intended patient population, but that racial and ethnic minority population groups still had lower participation rates than would be representative (with 77 percent of participants white, population average 72 percent).

In the commissioned analysis, among drugs that have been approved by the FDA in recent years, the committee found that women represented an average of 51 percent of participants between 2014 and 2021, ranging from 37 percent in 2014 to 54.8 percent in 2020 (data for 2021 are partial only). However, women's representation varies greatly by disease type. Prior to 2021, women represented greater than 50 percent of trial participants over at least 5 years in the areas of ophthalmology, gastroenterology, and endocrinology/metabolism/bone. However, women represent less than 50 percent of trial participants over at least 5 years in the areas of cardiovascular disease and infectious disease (see Figure 2-2). It is important to note that the participation of women may be driven by diseases and conditions that disproportionately affect women, such as osteoporosis and irritable bowel syndrome. While the committee did not examine clinical trial enrollment by specific disease burden, it is important to note that matching disease burden with trial representation is ideal, and therefore, 50 percent may not be the accurate threshold by which to measure women's participation in clinical trials and clinical research.

The committee also found similarly positive trends in clinical research participation of women in NIH-sponsored trials. The committee found that participation of women has been steadily increasing from 2013 to 2018 for which data are available (no data were reported in 2015, but reporting requirements changed in FY 2016, resulting in an increase in participants reported across NIH institutes and centers). Across all NIH institutes and centers, mean representation of women in clinical research was 44.3 percent in 2013, 47.2 percent in 2014, 54.1 percent in 2016, 47.9 percent in 2017, and 52.4 percent in 2018 (on average 22.1 million participants were included in NIH-funded trials during each of these annual reporting periods).

As shown in Figure 2-3, among the top 10 largest institutes/centers by research enrollment (which represent 89.7 percent of enrollment across all institutes/centers), women make up at least 50 percent of participants in clinical

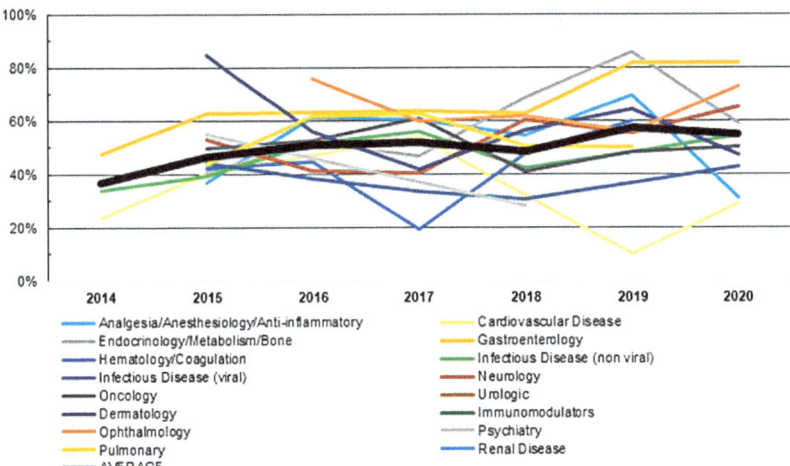

FIGURE 2-2 Average percent of women in trials by year of FDA approval and therapeutic area (n = 287).
SOURCE: Analysis of FDA Drug Trials Snapshots as of May 2021.

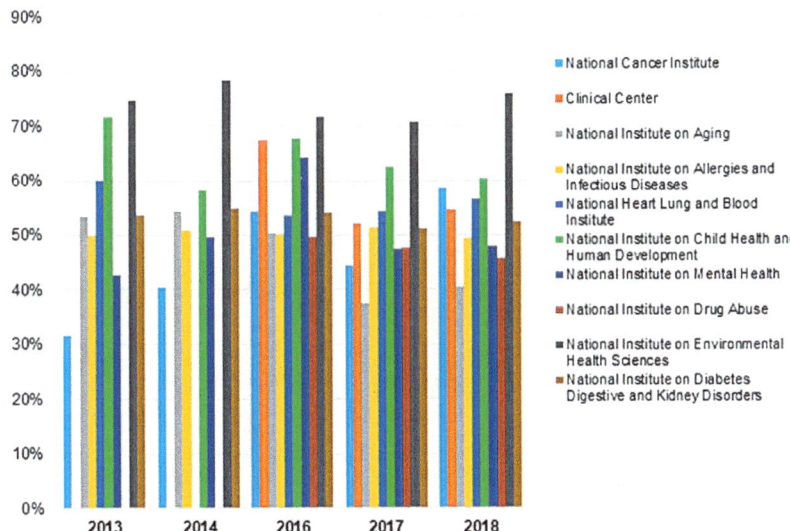

FIGURE 2-3 Participation of women in clinical trials supported by NIH institutes (top 10 institutes/centers by 2018 enrollment).

research supported by the National Institute of Environmental Health Sciences, the National Institute of Diabetes and Digestive and Kidney Diseases, and the National Institute of Child Health and Human Development across all years of reporting, and at least 50 percent of participants in at least 3 years of reporting in clinical research supported by the National Institute on Aging, the Clinical Center, the National Institute of Allergy and Infectious Diseases, and the National Heart, Lung, and Blood Institute. Across all 5 years of reporting, women never exceeded 50 percent of participants in clinical research supported by the National Institute on Minority Health and Health Disparities, the National Institute on Alcohol Abuse and Alcoholism, and the National Library of Medicine. Thus, representation of women, particularly white women, has improved in clinical research over the past decade; however, the evidence specifically on pregnant and lactating individuals, sexual- and gender-minority populations, and racial and ethnic subgroups of women is lacking (see the Racial and Ethnic Diversity in Clinical Trials section).[1]

Participation of Pregnant and Lactating Individuals in Clinical Trials

Nearly 4 million persons in the United States give birth every year, and 3 out of 4 gestational parents start out breastfeeding (Hamilton et al., 2021; HHS, 2022b). During pregnancy and lactation, greater than 90 percent of these individuals take at least one medication, either to treat pregnancy-related complications or to treat ongoing medical issues (NIH, 2018b). However, pregnant and lactating persons are often excluded from clinical trials and clinical research that could help them and provide better clarity on the risks and benefits of taking prescribed medications during pregnancy or while lactating. Very few drugs are approved for use during pregnancy, and most drug labels have little data on pregnancy to inform prescribing decisions (Blehar et al., 2013).

Despite federal initiatives to address this problem, pregnant and lactating individuals remain drastically underrepresented in clinical trials (Blehar et al., 2013). In a study of six clinical trial registries, just 0.32 percent of all active registered trials were focused on pregnant individuals (Scaffidi et al., 2017). Additionally, in a review of 338 Phase 3 and 4 NIH-funded actively recruiting studies in ClinicalTrials.gov, 68 percent explicitly excluded pregnant women and 47.3 percent excluded lactating women (Spong and Bianchi, 2018). Another review found that of 558 industry-sponsored studies, only 1 percent were designed for pregnant women and 95 percent excluded pregnant women (Shields and Lyerly, 2013).

One of the main reasons driving the continued lack of trials on pregnant and lactating individuals is the health risk posed to the survival of pregnant individu-

---

[1] Unfortunately, the data the committee used for this analysis were not disaggregated by sex and race/ethnicity, so an analysis on the status of the participation of racial and ethnic subgroups of women broken down by different racial and ethnic groups is not included.

als and their offspring. This is exacerbated by the highly publicized cases where drug trials ended with tragic results, such as with thalidomide. Following some of these trials, policies were passed in 1977 that effectively excluded pregnant women from clinical trials (Becker, 2021). The FDA walked back this broad ban in the 1990s, largely thanks to AIDS activists in the 1990s, who argued the policy effectively limited their access to life-saving drugs (Brick-Hezeau, 2019). However, pregnant individuals are still largely excluded from clinical trials beyond preclinical safety assessments (NIH, 2019). The potential risk to fetuses and legal consequences of injury to children who were exposed in utero present ethical and legal considerations for industry. In addition to safety concerns, physiologic changes that occur during pregnancy and while lactating can affect drug metabolism, leading to increased complexity in a clinical trial.

A recent example of the complexity of clinical trial design with pregnant and lactating persons is the COVID-19 vaccine trials, which excluded pregnant and lactating individuals from participating. Pregnant and lactating individuals were excluded from initial COVID-19 vaccine trials for safety concerns and because including them could have complicated and potentially delayed the use of vaccines for the broader population (Szabo, 2022). In addition, because little information on the use of mRNA vaccines in pregnant and lactating individuals existed, the FDA required drug companies to complete developmental and reproductive toxicity studies before testing in pregnant people (FDA, 2011b). These initial toxicity studies take 5 to 6 months to complete, meaning they were not completed for COVID-19 vaccine trials until late 2020.

However, the exclusion of pregnant and lactating persons from COVID-19 trials was made without any evidence suggesting that vaccines are teratogenic and without any evidence that they are transmitted to breast milk, leaving many without the necessary data to make an informed decision (Van Spall, 2021). This has led to the spread of misinformation on the impacts of COVID-19 vaccines on pregnancy and vaccine hesitancy for a high-risk group (Skirrow et al., 2022).

To address this continued challenge, the 21st Century Cures Act (P.L. 144-255) established the Task Force on Research Specific to Pregnant Women and Lactating Women (PRGLAC). PRGLAC was charged with "providing advice and guidance to the Secretary of Health and Human Services on activities related to identifying and addressing gaps in knowledge and research on safe and effective therapies for pregnant women and lactating women." The PRGLAC report, issued in 2018, includes a list of recommendations to address this issue (see Box 2-4) (NIH, 2018b). Since the report was published, the NIH has proposed an implementation plan for carrying out all of the recommendations in the report and calls on multiple stakeholders, including government, industry, clinicians, and women, to each do their part in carrying out these implementation steps (Byrne et al., 2020).

The committee recognizes that the inclusion of pregnant and lactating individuals in such trials may require special considerations, including medical

BOX 2-4
Task Force on Research Specific to Pregnant Women and Lactating Women Report Recommendations

1. Include and integrate pregnant women and lactating women in the clinical research agenda
2. Increase the quantity, quality, and timeliness of research on safety and efficacy of therapeutic products used by pregnant women and lactating women
3. Expand the workforce of clinicians and research investigators with expertise in obstetric and lactation pharmacology and therapeutics
4. Remove regulatory barriers to research in pregnant women
5. Create a public awareness campaign to engage the public and health care providers in research on pregnant women and lactating women
6. Develop and implement evidence-based communication strategies with health care providers on information relevant to research on pregnant women and lactating women
7. Develop separate programs to study therapeutic products used off-patent in pregnant women and lactating women using the National Institute of Health (NIH) Best Pharmaceuticals for Children Act (BPCA) as a model
8. Reduce liability to facilitate an evidence base for new therapeutic products that may be used by women who are or may become pregnant and by lactating women
9. Implement a proactive approach to protocol development and study design to include pregnant women and lactating women in clinical research
10. Develop programs to drive discovery and development of therapeutics and new therapeutic products for conditions specific to pregnant women and lactating women
11. Utilize and improve existing resources for data to inform the evidence and provide a foundation for research on pregnant women and lactating women
12. Leverage established and support new infrastructures/collaborations to perform research in pregnant women and lactating women
13. Optimize registries for pregnancy and lactation
14. The Department of Health and Human Services Secretary should consider exercising the authority provided in law to extend the PRGLAC Task Force when its charter expires in March 2019
15. Establish an Advisory Committee to monitor and report on implementation of recommendations, updating regulations, and guidance, as applicable, regarding the inclusion of pregnant women and lactating women in clinical research

SOURCE: NIH, 2018b.

clearance of the participants, specialized informed consent, and accelerated completion of reproductive safety and toxicology data on the drug or device under study. While acknowledging the extra steps that are required, examples such as COVID-19 vaccine trials highlight the clinical and scientific grounds where such inclusion can be essential and lifesaving.

Participation of Sex, Sexual and Gender Minority Populations in Clinical Trials and Clinical Research

Sex, sexual and gender minority (SGM) populations experience significant health disparities compared with their cisgender, heterosexual peers, including higher rates of cardiovascular disease, myocardial infarction, some cancers, and HIV. They are more likely to experience violence and trauma, and they report greater anxiety, depression, and suicidality (NASEM, 2020b); which may contribute to alterations in physiology, neurobiology, and immunity (van der Kolk, 2014). Additionally, there are unique considerations related to the biological effects of gender-affirming care and medical or surgical interventions that need to be explored (Jones et al., 2020).). Given the known health disparities and biological considerations, it is important for SGM populations to be included in clinical trials and clinical research.

The committee's search of the literature did not find much on the status of SGM participation in clinical trials and clinical research.[2] There is a literature base on the barriers and facilitators to SGM participation in clinical trials and research, but the committee could find only one analysis on the participation of SGM populations in clinical trials. In a manual analysis of ClinicalTrials.gov by Chen et al. (2019), researchers found a clear increase in transgender-recruiting trials over time, from zero reported trials in 2013, gradually increasing each year up to nearly 70 in 2018.

Part of the reason the analysis of SGM individuals in clinical trials is unknown is because clinical trial registries, such as ClinicalTrials.gov, define *Male, Female,* and *All* as structured information for the gender requirement entry and do not collect information about sexual orientation. To specify further that the trial is recruiting SGM individuals, that information is included in the inclusion criteria section (Chen et al., 2019). Due to the lack of routinely collected data on SGM in research, the NIH SGM research working group has in its strategic plan to (1) expand the knowledge base of SGM health and well-being through NIH-supported research; (2) remove barriers to planning, conducting, and reporting NIH-supported research about SGM health and well-being; (3) strengthen the

---

[2] The committee uses the NIH's definition of sexual and gender minority (SGM), which as utilized includes lesbian, gay, bisexual, and transgender people (LGBT) as well as those whose sexual orientation, gender identity and expressions, or reproductive development varies from traditional, societal, cultural, or physiological norms.

community of researchers and scholars who conduct research relevant to SGM health and well-being; and (4) evaluate progress on advancing SGM research.

Some research initiatives have focused specifically on SGM populations due to the lack of research data in this population. One such initiative is PRIDEnet, which enrolled 13,244 SGM people and for which researchers can propose studies on data from PRIDE, or request new studies with PRIDE participants (PCORI, 2019).

The NIH 2021–2025 strategic plan to advance research on the health and well-being of SGM populations reports that SGM are a health-disparities population, proposes to provide support of new investigators to build a strong SGM workforce, and will increase projects related to SGM health (NIH, 2021b). That being said, in many cases, sexual orientation and gender identity data are not collected in research, and when they are collected they are not always done so in a standardized way.

Racial and Ethnic Diversity in Clinical Trials

Numerous studies have highlighted the lack of diversity by race and ethnicity and the lack of reporting of these demographic characteristics. Here, the committee highlights several specific studies that provide illustrative evidence of underrepresentation of specific groups in clinical trials in the past two decades.

A 2004 analysis of cancer clinical trials found that Hispanic and Black patients were 28 percent and 29 percent, respectively, less likely to be enrolled than white patients after adjustment for disease incidence, age, and other factors. The difference was largest in lung cancer, where Black patients were 39 percent and Hispanic patients were 53 percent less likely to be enrolled than white patients.

A recent study of 230 vaccine trials from 2011 to 2020 indicated that white participants tend to be overrepresented, while Black and other racial or ethnic minority participants tended to be underrepresented. The enrollment of Asian individuals was perhaps approximate to the U.S. Census estimates (Flores et al., 2021). A report on the diversity of mRNA vaccine trials for COVID-19 by the Kaiser Family Foundation has found a relatively higher share of white participants in both trials compared with the U.S. population, resulting in relative underrepresentation of Black and Asian participants. However, the participation of Hispanic adults exceeded the share of Hispanic adults in the U.S. population (Artiga et al., 2021). These results, however, originated from trial sites within (76.7 percent) and outside of the United States (notably Europe and Latin America), which may explain some of the relative overrepresentation of white and Hispanic participants.

Even recently completed trials have failed to include enrollment consistent with the distribution of disease across the population—a Phase 2 trial of crenezumab in Alzheimer's disease with 360 participants across 83 sites in 6 countries reported 97.5 percent of participants being white, and only 2.8 percent of all participants being Hispanic, for example (Genetech Inc., 2020). However, African Americans are about two times more likely than white Americans to develop Alzheimer's disease and other dementias, and Hispanics are about one and half times as likely as white Americans to develop Alzheimer's disease and other dementias (AIM, 2020).

In the commissioned analysis of the FDA Drug Trials Snapshots data, the committee found that among approved drugs, participation of white patients has ranged from 84 percent in 2014 to 73.7 percent in 2020, indicating a relatively consistent decrease in the share of white participants in trials resulting in FDA approval during this period (2021 data are yet incomplete). However, the increasing inclusion of data from international trial sites in FDA drug approvals means that this trend may not represent increases in U.S.-residing racial and ethnic minority population groups. Similarly, these data do not speak to whether the demographic distribution in a trial is reflective of the demographic distribution of those affected by the condition being studied.

NIH reporting on ethnicity and race are not always consistent (see Table 2-2), but results show a relatively stable trend of proportion of participants across racial and ethnic groups in clinical trials with the weighted average of white participants among the top 10 institutes ranging from 51.8 percent in 2013 to 60.6 percent in 2018 (this trend mirrors that of all NIH-sponsored trials, as shown in Figure 2-2). Interpretation of these data is limited because the demographic characteristics of the condition or disease under study is not included.

TABLE 2-2 Demographics of Participants in Trials Supported by NIH Centers and Institutes

	2013 (%)	2014 (%)	2016 (%)	2017 (%)	2018 (%)
Female	44.3	47.2	54.1	47.9	52.4
American Indian	2.1	1.3	0.8	0.7	1.0
Asian	15.1	17.2	8.4	26.4	7.8
Black/African American	12.2	14.3	10.0	10.8	13.5
Native Hawaiian/Pacific Islander	0.3	0.3	0.6	0.1	0.2
White	52.9	49.5	49.6	49.9	60.0
More than 1 race	1.1	1.1	2.0	1.9	2.3
Unknown race	1.1	1.1	2.0	1.9	2.3
Hispanic	9.8	8.1	10.8	6.7	8.5
Non-Hispanic	86.1	89.6	62.6	81.8	76.2
Unknown ethnicity	4.1	2.3	22.4	9.8	12.0
Sum of all races	84.7	84.8	73.5	91.8	87.2
Sum of all ethnicities	100.0	100.0	95.8	98.3	96.7

NOTE: The full analysis is available in Appendix B.

To examine the breakdown of race and ethnicity by center, the committee examined clinical research participation for each reported racial and ethnic group sponsored by the top 10 largest NIH institutes. The data on participation were collected, aggregated, and analyzed from biennial and triennial reports provided by each of these institutes. Overall, it appears that demographic trends in NIH-funded clinical research have not changed much over the years (see Appendix B for details). However, these trends can vary widely by institute. For example, the National Institute of Allergy and Infectious Diseases reported participation of African American/Black greater than 25 percent in all years examined, while the National Cancer Institute reported just 10.5 percent at most.

3

Policies to Improve Clinical Trial and Research Diversity: History and Future Directions

The analysis draws substantially from the research paper by Dr. Laura Bothwell, Ph.D., and Aaron Kesselheim, M.D., J.D., M.P.H, which was commissioned for this study. The full research paper can be found online at: nap.nationalacademies.org

In this chapter, the committee describes major federal policies designed to improve the inclusion of underrepresented and excluded populations in clinical trials and analyze the benefits and limitations of these policies with the aim of improving them. The history of trial diversity policies is deeply embedded in the broader historical context of work toward equity and inclusion. Given the statement of task, the committee decided it was most appropriate to get a scope of what political action has been taken to include more diversity in clinical trials and clinical research. However, the committee would like to acknowledge that there have been many federal policies throughout history that have contributed to racial and ethnic groups being excluded, such as census policies related to race and ethnicity (Pratt et al., 2010). Although out of scope for this report to cover in great detail, the committee deemed it important to acknowledge that federal policies have historically both increased inclusion and exclusion.

EARLY HISTORY

Race and Ethnicity

The Civil Rights Act of 1964 (P.L. 88-352) was perhaps the earliest occasion when legislators or regulators set policies on racial diversity in clinical research. In compliance with the law, in 1965, National Institutes of Health (NIH) General

Clinical Research Centers added new notices to grant applications warning that racial discrimination was illegal. Eventually, all domestic U.S. grant applicants to the Department of Health and Human Services (HHS) had to file with the HHS Office for Civil Rights an assurance of compliance with Title 6 of the act, which prohibits discrimination based on race, color, religion, national origin, or sex in services and establishments that receive federal funding, including hospitals and medical facilities.

Since that time, enforcement of Title 6 has been partial and inconsistent, and racial and ethnic minority populations groups have continued to experience inadequate treatment in clinical care and research at both federal and state levels (Yearby, 2014). National attention was drawn to problems of racism in research in 1972 with the revelation of the 40-year Tuskegee Syphilis Study conducted by the U.S. Public Health Service observing the progression of syphilis among untreated low-income African American men long after treatment had become available (Brandt, 1978). In response, Congress passed the National Research Act of 1974 (P.L. 93-348), which established the National Commission for the Protection of Human Subjects of Biomedical and Behavioral Research (Vargesson, 2015). The commission published the *Belmont Report: Ethical Principles and Guidelines for the Protection of Human Subjects of Research*, which laid the groundwork of principles and guidelines for research involving human subjects, identifying three basic ethical principles for human subject experimentation: respect for persons, beneficence, and justice. The report pointed out that "the selection of research subjects needs to be scrutinized to determine whether some classes (e.g., welfare patients, particular racial and ethnic minority population groups, or persons confined to institutions) are being systematically selected simply because of their easy availability, their compromised position, or their manipulability, rather than for reasons directly related to the problem being studied" (National Commission for the Protection of Human Subjects of Biomedical and Behavioral Research, 1979).

Sex and Gender

In the mid-20th century, alongside growing awareness of the value of protecting vulnerable populations, many began to draw attention to a long-held bias in the field of clinical research: the "male norm," as later summarized in a 1994 report by the Institute of Medicine (1994). Healthy, young, or middle-aged males, frequently who were white, were considered to be the "norm" study population; by contrast, females were thought to confound trial results with their fluctuating hormone levels and reproductive potentials (IOM, 2001; Pinn, 2003). When news broke from Europe and Canada in the early 1960s that widespread maternal exposure to the sedative thalidomide during pregnancy led to fetal death and birth defects, policy makers took the stance that pregnant women were a "vulnerable population" who should be shielded from the potential reproductive adverse effects of drug exposures in trials (Vargesson, 2015). In response to the tragedy, the

U.S. Congress passed the Kefauver-Harris Amendment in 1962 (P.L. 87-781) to strengthen the authority of the Food and Drug Administration (FDA) in overseeing drug development and pre-market evaluation. Some years later, in 1977, the FDA created a guideline, "General Considerations for the Clinical Evaluations of Drugs," that banned women of childbearing potential from Phase 1 and early Phase 2 trials, except for life-threatening conditions (FDA, 1977). The policy strictly excluded women who used contraception, who were single, or whose husbands had vasectomies (FDA, 1993). In 1979, the *Belmont Report* further stipulated that pregnant women should be considered vulnerable research subjects and should be protected at all costs (National Commission for the Protection of Human Subjects of Biomedical and Behavioral Research, 1979). The 1979 FDA Labeling Rule established the first classification system for identifying the risks prescription drugs posed to pregnant women, fetuses, and breastfeeding infants (FDA, 1979).

Age

The phrase "therapeutic orphan" was coined by Harry Shirkey, M.D., in 1963 to describe the lack of modern drug therapy targeted toward children (Shirkey, 1968). Most authors have attributed this state of affairs to the shortage of relevant drug research in children, as private-sector sponsors deemed the introduction of therapies targeting children to have little potential for profit (MacLeod, 2010). In 1978, the National Commission for the Protection of Human Subjects of Biomedical and Behavioral Research published a report on research involving children, discussing the fundamental ethical permissibility of pediatric research, particularly research not benefiting the child involved (NCPHS, 1978b). Prior to this report, philosophers and ethicists held opposing views: some repudiated any ethical justification for research with a healthy child (Ramsey, 1976), while others claimed that even children bear a certain obligation to benefit society, justifying a presumption of their consent to experiments of minimal risk (McCormick, 1974). The commission contended that children might be entered in research entailing more than minimal risk and promising no individual benefit when (1) the risk entailed represents "a minor increase over minimal risk," (2) the experience presented by the intervention is "reasonably commensurate with those inherent in the actual or expected medical, psychological or social situation" of the subject, and (3) the research is likely to yield generalizable knowledge about the subject's condition that is "of vital importance for the understanding and amelioration of the condition" from which that class of subjects suffers (Jonsen, 1978). These recommendations were later adopted by the Department of Health and Human Services, including the FDA, in its regulation titled "Additional Protections for Children Involved as Subjects in Research" in 1983 (HHS, 1983).

In addition to creating risk classifications for drugs taken by pregnant women and lactating mothers, the 1979 FDA Labeling Rule sought to improve the safety and efficacy of drugs intended for diverse ages by requiring labeling content under "Pediatric Use" and "Geriatric Use." Notably, the rule did not outline specific requirements for risk information provided under the pediatric and geriatric sections as it did for data on pregnant women and lactating mothers (FDA, 1979).

Diversity among Investigators

The Office of Minority Programs (OMP) was established in the NIH Office of the Director in 1990 (NIMHD, 2022). Two years later, the OMP co-funded various projects, including training for faculty and students at all stages along the educational pipeline. It also funded a National Academy of Sciences study that focused on evaluating NIH training programs for underrepresented students (NRC, 2005). The study found that while many NIH programs were helpful in providing students research experience, funding, and mentoring, there was a sharp drop-off among "minority trainees" at the postdoctoral and junior faculty levels. The OMP eventually became the National Center on Minority Health and Health Disparities in 2000, and was redesignated as the National Institute of Minority Health and Health Disparities in 2010.

In addition to the OMP, individual institutes have their own ongoing initiatives. For example, the NIH Office of Diversity and Health Disparities (ODHD) within the National Institute on Drug Abuse was established more than 20 years ago and serves to strengthen a more diverse and robust extramural research workforce, attracting and retaining talented individuals from all backgrounds, and supporting research aimed at the NIH mission of reducing health disparities. Among its numerous endeavors, it provides funding to recruit and support high school, undergraduate, and graduate/clinical students, postdoctorates, and eligible investigators to work on an existing NIH-funded project in a particular area of interest (NIH, 2021d). This opportunity is also available to investigators who are or become disabled and need additional support to accommodate their disability to continue to work on the research project.

MODERN POLICIES

National Institutes of Health

The NIH is responsible for providing direction to research programs with goals to improve the health of the nation, and to that end, the NIH creates policies to improve the nation's well-being (NIH, 2021a). The NIH is the largest federal sponsor of clinical trials in the United States, devoting about $3 billion per year to funding trials (NIH, 2017c). Its stewardship over clinical trial policies has a substantial impact on the rigor, transparency, and effectiveness of the clinical trial enterprise (Hudson et al., 2016).

The first significant work toward inclusive clinical trial policies at the NIH emerged in response to the 1985 report of the U.S. Public Health Service Task Force on Women's Health Issues outlining how underrepresentation of women in clinical trials had led to suboptimal women's health care (Women's health, 1985). The task force recommended increased participation of women in clinical trials, including women of childbearing potential. It also recommended that research should emphasize diseases that are more prevalent in women (Liu and Dipietro Mager, 2016). In response to this report, the NIH adopted the Inclusion of Women and Minorities in Clinical Research policy in 1986 (NIH, 1987). The major goal of this policy was to ensure that research and clinical trials were designed to provide information about sex and race/ethnicity differences. Response to this policy was slow; guidance for its implementation was not developed until 1989 when a memorandum on inclusion announced that research solicitations should encourage the inclusion of women and minority population groups and stipulated that a rationale should be provided when women and minority population groups were excluded (NIH, 1989).

In 1990, the General Accounting Office (GAO), later known as the Government Accountability Office, a legislative branch agency that provides auditing, evaluation, and investigative services for the U.S. Congress, investigated the NIH implementation of the guidelines for the inclusion of women and minority population groups. In its report, the GAO revealed that the 1986 Inclusion of Women and Minorities in Clinical Research policy had been poorly communicated and inconsistently applied before the 1990 grant review cycle. The GAO identified two major limitations of the policy. The policy only pertained to extramural research conducted by investigators who had been awarded NIH grants, but not intramural research overseen by scientists employed by the federal government. In addition, the policy provided little incentive for researchers to analyze study results by gender (Nadel, 1990). As criticism mounted in response to the GAO report, the Congressional Women's Caucus took legislative action by passing a package of bills known collectively as the Women's Health Equity Act of 1990 (S. 2961, 101st Congress (1989–1990)). Responding to this new legislation, the NIH founded the Office of Research on Women's Health (ORWH) in the same year (P.L. 103-43). The ORWH helped the research community understand the importance of inclusion of women in clinical trials by monitoring and promoting NIH-wide efforts to ensure the representation of women and by prioritizing diseases, disorders, and conditions that primarily affect women. The ORWH also supports initiatives to advance women in biomedical careers and ensures that women are included in clinical research funded by the NIH (P.L. 103-43).

The establishment of these offices and lessons learned from the original inclusion policy contributed to the development of the NIH Revitalization Act of 1993 (P.L. 103-43), which became an updated version of the original inclusion policy but also provided additional guidance on the inclusion and reporting and analysis of sex/gender and racial/ethnic differences in intervention effects for

NIH-defined Phase 3 clinical trials (Night, 2009). The act emphasized that the NIH should ensure that women and minority population groups be included in all clinical research, that Phase 3 clinical trials had sufficient numbers of participants to allow for analysis, that populations were not to be excluded from trials due to cost, and that the NIH must maintain outreach efforts to include women and minority population groups in clinical studies. The law was designed to ensure that clinical research determines whether an intervention differently affects men, women, or members of a minority population (Liu and Dipietro Mager, 2016). Scientists at the time largely supported the act, and it sparked discussions about the importance of appropriate trial design and subsequent subgroup analyses. After implementation of the act, women and minority population groups were increasingly included in clinical trials (Boissel et al., 1995; Freedman et al., 1995). Currently, females make up 49 percent of subjects in NIH-funded clinical trials (Blehar et al., 2013). Under the act, the Office of Minority Programs also changed its name to the Office of Minority Health Research (OMHR). At this point, the OMHR did not have grant-funding authority.

Key gaps remained regarding inclusivity of NIH-funded trials. A study comparing the ethnic distribution of patients enrolled in trials funded by the National Cancer Institute in 2000 through 2002 with those enrolled in 1996 through 1998 found that the proportion of minority trial participants did not change significantly and that the proportion of participants who were Black had declined. After adjusting for age, cancer type, and sex, patients enrolled in 2000 through 2002 were 24 percent less likely to be Black than those enrolled in 1996 through 1998 (Murthy et al., 2004). Ten years into the NIH Revitalization Act's implementation, another GAO report found that although women were taking part in clinical studies in greater numbers than men and more funding was available for studying diseases that disproportionately affected women, only a small fraction of publications based on NIH-funded research reported findings stratified by sex (Helmuth, 2000). Twenty years post–NIH Revitalization Act, another study concluded that minority population groups remained disproportionally underrepresented in cancer clinical trial enrollments in 2014. In addition to persistent barriers for minority participation in cancer clinical trials, the study reported a dearth of cancer clinical trials that focus primarily on racial/ethnic minority populations, as well as a lack of usable trial data about racial/ethnic minority populations (Chen Jr. et al., 2014). The analysis of NIH-funded trials commissioned by the Committee on Women in Science, Engineering, and Medicine showed that demographic trends in NIH-funded clinical research have not changed much over the years (see Appendix B) and that these trends can vary widely by institute.

In 1998, the NIH Guide for Grants and Contracts published guidelines for including children in research supported by the NIH, unless there were scientific or ethical reasons not to include them (NIH, 1998). The goal of the policy was to obtain appropriate data on treatment outcomes in children. This policy applied to all initial applications/proposals and intramural projects submitted to the

NIH, and it provoked discussions among investigators and ethicists surrounding the ethical dilemma of balancing improving access and recruitment of children in clinical trials with the need to protect this vulnerable population (Glantz, 1998; Kopelman, 2000; Tauer, 2002). The impact of this guideline seemed to lag behind those targeting women and minority populations' enrollment. A survey was conducted in 2008 to assess NIH Scientific Review Group (SRG) members' experiences with and attitudes about the NIH inclusion guidelines for women and minority population groups and children, released in 1994 and 1998, respectively. While about half of the SRG members surveyed agreed that the inclusion guidelines resulted in an increase in the number of underrepresented and excluded populations enrolled in clinical research, less than one-third responded that the guidelines expanded the inclusion of children (Taylor, 2008).

In 2000, with the passage of the Minority Health and Health Disparities Research and Education Act (P.L. 106-525), the office became the National Center on Minority Health and Health Disparities (NCMHD). The act gave NCMHD the authority to fund grants and called for the development of a comprehensive NIH strategic research plan and budget for health disparities research. The center was again redesignated as the National Institute on Minority Health and Health Disparities in 2010 with the passage of the Patient Protection and Affordable Care Act (P.L. 111-148), or ACA (NIH, 2010). The office gained authority to plan, review, coordinate, and evaluate the minority health and health disparities research and activities conducted and supported by the NIH institutes and centers (Kneipp et al., 2018).

In 2001, the NIH policy and guidelines on the Inclusion of Women and Minorities as Subjects in Clinical Research from 1986 were updated (NIH, 2001a). The original purpose of the 1986 policy was to ensure the inclusion of women and minority groups in NIH-funded clinical research and that these research findings should be generalizable to a broad population (IOM, 1994). However, the policy lacked a clear definition of clinical research and did not require specific analyses by racial groups to be included when reporting population data. Thus, the updates provided guidance on clarifying the definitions of racial and ethnic categories and reporting analyses of sex and racial minority population groups in clinical trials (Nours, 2021). These updates included the Office of Management and Budget (OMB) Directive's racial and ethnic categories that are to be used to monitor population data for clinical trials. Though the directive claimed that "the categories in this classification are social political constructs and should not be interpreted as being scientific or anthropological in nature," scholars argued that standards reflected an important step in moving beyond a simplistic concept of race and its impact on health and provided state and federal public health agencies with an important opportunity to collect, tabulate, and analyze data on program participation and community health that more accurately reflected the racial and ethnic nuances of contemporary American society (Friedman et al., 2000; Hattam, 2005). According to a 2015 GAO report, however, the reporting of the racial/ethnic composition of study participants did not improve since 2004 (GAO, 2015).

In 2009, the NIH commissioned the Institute of Medicine to conduct a study on the health of lesbian, gay, bisexual, and transgender (LGBT) individuals. The resulting report, *The Health of Lesbian, Gay, Bisexual, and Transgender People: Building a Foundation for Better Understanding*, concluded that major knowledge gaps exist in the health needs of LGBT people and urged the NIH to support additional research (IOM, 2011). The NIH LGBT Research Coordinating Committee was established to develop and coordinate the NIH's LGBT research and training, expand knowledge of LGBT health, and improve methods to reach these populations through specific trial networks such as the Adolescent Medicine Trials Network for HIV/AIDS Interventions, HIV Prevention Trials Network, HIV Vaccine Trials Network, and Microbicide Trials Network (NIH, 2015a). In 2015, the Sexual and Gender Minority Research Office was created to coordinate and support sexual and gender minority research activities across the NIH.

To support transgender inclusion, the trial networks adopted a two-step method in data collection forms, separating birth sex and gender identity into two variables (Sausa et al., 2009). In addition, the trial networks updated protocol design with language for transgender inclusion, implemented staff training for cultural sensitivity, consulted with transgender individuals, and conducted new research on transgender individuals (Siskind et al., 2016).

In his 2015 State of the Union address, President Obama announced the Precision Medicine Initiative. In response to this initiative, the NIH created an ethnically diverse research cohort amounting to 1 million or more Americans who had agreed to have their clinical data tracked for research purposes (Collins and Varmus, 2015). This effort was accompanied by workshops hosted by the NIH that examined the reproducibility and transparency of clinical research and aimed to maximize cohort diversity, inclusion, and attention to health disparities (ACD, 2015; NIH, 2015b, 2015c). To further catalyze diversity in research, analysts suggested that the NIH should be empowered to set and enforce recruitment of diverse research populations by race and ethnicity as the default and require scientific justification for limited or selected study population enrollment, similar to what had been created for sex balance (Clayton and Collins, 2014; Oh et al., 2015).

The NIH revisited its policies on age in response to the passage of Section 2038(H) of the 21st Century Cures Act in 2016 (P.L. 114–255). This act instructed the NIH to hold a workshop accounting for differences across the lifespan, publish guidelines addressing consideration of age in clinical research, and ensure that researchers conducting applicable Phase 3 clinical trials report results of analyses by sex/gender at ClinicalTrials.gov (Nours, 2021). As a result of these efforts, the Inclusion Across the Lifespan (IAL) policy was created. This policy required that NIH-funded studies include individuals of all ages (including older adults and children) in clinical trials unless age-based exclusions are scientifically or ethically justified. The policy outlined when certain age groups may be excluded and noted that grantees are required to annually report on the age at enrollment of their participants along with sex/gender, race, and ethnicity (Nanna et al., 2020; NIH, 2017a, 2017b, 2020a).

The IAL policy is still in its nascent stages, and more data are needed to assess its impact (the policy went into effect in January 2019) (Nanna et al., 2020). Further policy work may also be warranted, as solely extending the age of eligibility for clinical trials is insufficient to make a study truly representative of the general population because social factors such as socioeconomic status may influence access to trials by marginalized groups (Lauer, 2020).

An IAL workshop held in 2017 resulted in several publications related to people with disabilities, including a *JAMA* article reporting that a present-day review of 338 Phase 3 and 4 NIH-funded actively recruiting studies in ClinicalTrials.gov found that most of the trials did not mention individuals with disabilities in either the inclusion or exclusion criteria (greater than 90 percent did not mention physical disabilities and greater than 80 percent did not mention intellectual disabilities) (Lockett, 2017; Spong and Bianchi, 2018). Explicit exclusion was mentioned in 12.4 percent of the studies for those with intellectual or developmental disabilities (including criteria based on IQ, defined intellectual disability, or cognitive impairment). Explicit exclusion was mentioned in 1.8 percent of studies for those with physical disabilities (including inability to ambulate, extreme immobility, and paraplegia) (Spong and Bianchi, 2018). Further, there are non-explicit barriers to trial participation for people with disabilities, as those with cognitive impairment may be limited by lack of ability to comply with the study protocol or procedures, and individuals with physical disabilities can face limited access to study facilities or face challenges with physiological measurements.

The Sex as a Biological Variable (SABV) policy, which was passed in January 2016, plays an important role in consideration of preclinical research and the design of clinical trials. It established the expectation not only that gender be considered when volunteers sign up for a study but also that investigators balance the proportion of males and females in preclinical investigations from the earliest stages of study design (Arnegard et al., 2020). The policy requires researchers to take sex into account when creating research questions, designing experiments, analyzing data, and reporting results (Nours, 2021). In the 6 years since the NIH enacted SABV, progress has been made (Clayton, 2021). A survey of NIH study section members revealed growing favorability toward the policy, despite some unsupportive perspectives. The number of grant applications that appropriately consider SABV also has increased (Woitowich and Woodruff, 2019).

Regarding diversity among investigators, the NIH Advisory Committee to the Director Working Group on Diversity was formed in 2013 in response to the Working Group on Diversity in the Biomedical Research Workforce (WGDBRW) recommendations. The WGDBRW includes a subgroup on individuals with disabilities that focuses on systematically identifying data, strategies, and experiences of individuals with disabilities in the scientific workforce to address the multiple barriers they face. In 2017, the WGDBRW established a second subgroup, the Diversity Program Consortium, which supports numerous initiatives designed to build infrastructure leading to diversity, research mentorship for diverse scientists, and awards and resource support (NIH, 2013).

Despite these efforts, the 2021 *Women, Minorities, and Persons with Disabilities in Science and Engineering* report found that even though the share of science and engineering degrees awarded to underrepresented populations increased over the past decade, several disparities remained. Scientists and engineers with disabilities have an unemployment rate much greater than their peers, and even greater than that of the U.S. general labor force. Female scientists and engineers have lower median salaries than do their male counterparts in most broad occupational groups. Underrepresented populations also hold a small (8.9 percent) share of academic positions, which is considerably lower than their share of the population (NSF, 2021).

Enforcement and Accountability

To ensure the success of NIH inclusion policies, internal monitoring systems include offices, working groups, and committees established across the NIH. An example of a committee used to monitor the progress of NIH policies on the inclusion of underrepresented and excluded populations across the lifespan in clinical research is the Inclusion Governance Committee, which is responsible for monitoring NIH extramural grants and ensuring diversity reporting (Nours, 2021). The NIH also seeks information and advice from the public and hosts workshops that provide researchers with evidence-based approaches in meeting these policies. For example, the Inclusion Across the Lifespan-II workshop provided researchers information about the inclusion of pediatric and older populations in clinical studies in meeting the IAL policy (NIH, 2020a).

Accountability to inclusion policies occurs differently for intramural and extramural clinical research. Intramurally, monitoring for adherence to these policies occurs primarily at the scientific or chief director level. Extramurally, researchers applying for NIH grants must justify their study populations as part of the process to be considered for funding. Extramural researchers must also work with NIH staff to resolve any issues concerning lack of inclusion of certain populations prior to grant approval. Progress reports on a study's development are monitored by NIH program officers to ensure that all principal investigators meet an acceptable threshold for the number of participants and inclusion criteria in the study's population. Phase 3 clinical trials are required to report results of sex/gender and race/ethnicity data into ClinicalTrials.gov so that this information can be monitored (Nours, 2021).

The NIH also requires that funded researchers submit a Research Performance Progress Report, or RPPR, annually that asks grantees about their current accomplishments for the project, upcoming plans, and significant changes regarding human or animal subjects (NIH, 2018a). This information is entered into eRA Commons and is used for accessing and sharing information over the life of a study (NIH, 2016). These progress reports must be approved by NIH for continued funding. NIH then externally reports their inclusion data in a format that is disaggregated by Research, Condition, and Disease Categorization (RCDC) categories. These RCDC data can be found on the NIH Research Portfolio Online Reporting Tools (RePORT) website (Nours, 2021).

Throughout the history of NIH policies, adaptation has been critical, as some policies have not been sufficient to encourage scientists to broaden their study inclusion criteria. Although current policies are encouraging, underrepresented and excluded populations are still underrepresented in clinical trials of some diseases, such as cardiovascular disease, hepatitis, digestive diseases, HIV/AIDS, Alzheimer's disease, and chronic kidney disease. Improper analyses and disaggregated data in publications exist due to the lack of inclusion in clinical research, impeding the generalizability of scientific findings to the broader population (Nours, 2021). Therefore, further adaptation is needed to adequately diversify clinical trial participation. Investigator bias must also be addressed. In a 2018 study to evaluate compliance with inclusion and assessment of women and racial and ethnic minority population groups in randomized controlled trials, it was found that both male and female researchers perform equally poorly during analysis and reporting of women in clinical studies, and both male and female participants show the same amount of gender bias in decision making (Geller et al., 2018).

Much work also remains to achieve compliance with existing policies. Organizations such as the ORWH, which monitor compliance, are crucial. The ORWH has created resources such as the Inclusion Outreach Toolkit to help principal investigators fulfill their responsibility to conduct inclusive research (Mistretta and Mistretta, 2016; Nours, 2021). Furthermore, the NIH created three free e-learning courses as well as a high-level quarterly publication called *Women's Health In Focus at NIH* to raise awareness of the health of women and marginalized populations (Nours, 2021). Strengthening ORWH and other institutional accountability mechanisms could likely improve achieving inclusivity objectives.

Food and Drug Administration

The FDA has been working for decades to ensure that people of different ages, races, ethnic groups, and genders are included in clinical trials. The official stance of the FDA is that clinical trial participants should be representative of the patients who will ultimately use the medical products that the FDA evaluates, because people of different ages, ethnicities, or races can react differently to medical products for a variety of reasons (NIH, 2020b). The agency has primarily promoted diversity by publishing guidelines that inform sponsors and drug manufactures of the FDA's current thinking and regulatory interpretations (FDA, 2021a).

In 1985, the FDA introduced "Content and Format of a New Drug Application" (21 CFR 314.50 (d)(5)(v)), its first guidance on analyzing specific subgroups such as pediatric, geriatric, and renal failure patients to evaluate whether dosing modifications were necessary in these populations (FDA, 1985). The inclusion of renal failure patients could be considered early progress for individuals with disabilities. This regulation did not include gender and race as subgroups.

An early breakthrough for gender inclusion came following the work of the first HHS task force on women's health, established in 1983, which produced a 1985 report on women's health issues encouraging reexamination of extant policies excluding women of childbearing potential from clinical trial participation (HHS, 1985). The FDA responded with the 1987 publication of a guidance for industry, "Guideline for the Format and Context of the Nonclinical Pharmacology/Toxicology Section of an Application," which set an expectation that both sexes of animals should be used to provide valuable information in preclinical drug safety studies (FDA, 1987). In the following year, the FDA released "Guidance for the Format and Content of the Clinical and Statistical Section of an Application" in which it recommended analyzing data from clinical pharmacology studies for safety and efficacy by sex, race, and age (FDA, 1988). In addition, the FDA issued a 1989 guidance aimed at drugs used in the elderly that included "Guidelines for the Format and Content of the Clinical and Statistical Sections of an Application" (FDA, 1989). This guideline recommended the analysis of safety and efficacy data to determine the influence of demographic factors such as age and sex in Phase 2 or Phase 3 trials (the final two stages of clinical testing prior to drug approval).

Although the 1988 and 1989 guidance documents aimed to promote evaluation of drug effectiveness based on gender, a landmark GAO report in 1992 concluded that women were nonetheless being underrepresented in clinical trials and trial data were often not analyzed for differences in therapeutic response by sex (GAO, 1992). This report was prompted by a request from Congress based on studies in the medical literature that women tended to metabolize antihypertensive and cardiovascular drugs at a slower rate than men, and that drug interactions with female hormones and use of oral contraceptives could have caused different responses putting women at risk if the FDA approved drugs on the basis of clinical trials in which women were underrepresented (GAO, 1992; Tamargo et al., 2017). The GAO report found that for greater than 60 percent of drug trials, the representation of women in the trial population was less than the proportion of women in the population with the corresponding disease. The GAO concluded that the FDA had not issued adequate guidance for drug manufacturers to determine the extent and sufficiency of female representation in Phase 1 and 2 trials. For example, the FDA did not define the term *representative,* and drug manufacturers were uncertain of FDA expectations around that term.

While the 1992 GAO report did not evaluate the "appropriateness" of the FDA policy of excluding women of childbearing potential, in 1993 the FDA withdrew its restriction on the participation of women in early clinical trials (GAO, 1992). This retraction was believed to have been prompted by analyses of published clinical trials that showed that trials of aspirin or antianginal drugs had few or no women in them, which made it uncertain how they worked in women (FDA, 1993a; GAO, 1992). In addition, there had been concerns that the 1977 policy may have led to a general lack of participation of women in drug development studies (DiPietro and Liu, 2016). Concerns about the efficacy of drugs in women also arose at a time when the FDA and the scientific community were focusing the need for individualized treatment, and there had been a lack of specific studies of pharmacokinetics in women even when gender-related differences may be expected or important, such as differences due to menopause or the menstrual cycle, or oral contraceptive use, or differences based on body fat percentage, weight, or muscle mass. In addition, the 1977 policy had prevented the gathering of early information on drug response in women that could be used in the design of Phase 2 and 3 trials and may have delayed discovery of gender-based variation in drug effects (FDA, 1993). Earlier participation of women in clinical trials could have led to making appropriate gender-based adjustments in larger studies, such as doses based on weight rather than fixed doses. Still, the FDA did not require that women be included in trials (Wood, 2021). The agency merely stated that it would expect careful, gender-based characterization of drug effects, such as quantifying differences in dose-response and maximum size of effects. The FDA also recommended pharmacokinetic and pharmacodynamics screening in women as a tool to detect differences and analyses of safety and efficacy by sex.

In 1994, the FDA Office of Women's Health was established to guide the agency on policies for the inclusion of women in clinical trials (HHS, 1994). Within the same year, an Institute of Medicine report, *Women and Health Research: Ethical and Legal Issues of Including Women in Clinical Studies, Volume 2, Workshop and Commissioned Papers*, called attention to the forms of historical gender bias in the design and implementation of trials (IOM, 1994). Spurred by these concerns, Congress released a 1997 regulation, "FDMA Section 115: Clinical Investigations (b) Women and Minorities Regulation," that required the FDA and NIH to review and develop guidance on the inclusion of women and minorities in clinical trials.[1] To comply with this regulation, the FDA issued the Demographic Rule in 1998, revising the New Drug Applications (NDA) content to require safety and efficacy data to be presented by gender, age, and racial subgroups and dosage modifications to be identified for specific subgroups (FDA, 1998). This rule gave the FDA the authority to refuse any NDA that did not ana-

---

[1] Food and Drug Administration Modernization Act of 1997, S. 830, 105th Congress, November 21, 1997.

lyze safety and efficacy data appropriately. It also required data on participation in Investigational New Drug (IND) applications to be presented by sex, age, and race so that any potential deficiencies in the NDA submission could be identified. In addition, a 1999 FDA guidance recommended the use of population pharmacokinetics to help identify differences in drug safety and efficacy among population subgroups (FDA, 1999). In 2000, Congress passed a law titled Amendment to the Clinical Hold Regulations for Products Intended for Life-Threatening Diseases (21 CFR 312.42) that permitted the FDA to place clinical holds on IND studies if men or women were excluded due to reproductive potential from clinical trials on a serious or life-threatening illness.

Although these policies aimed to increase the inclusion of underrepresented and excluded populations in trials, there were important shortcomings. The 1998 Demographic Rule had the force of law but lacked specificity relative to previous guidance, as it did not include criteria to determine the number of women to be included. The guidance issued in 2000 also did not require the inclusion of any particular number of men or women. A 2001 GAO report that examined these policies found that around one-third of NDAs and 39 percent of IND documents failed to meet the requirements of the 1998 FDA regulation. Although the FDA had the authority to suspend research if women were excluded for their reproductive potential, it had never done so. In addition, the report found that women were only 22 percent of the participants in the small-scale safety trials in which dosage levels were set. There also was no management system to track the inclusion of women in trials or to monitor compliance with existing regulations. The FDA had no criteria to determine whether reviews of NDAs adequately addressed sex differences, and FDA medical officers had not been required to discuss sex differences in their own reviews of NDAs. Thus, the FDA lacked tools to enforce its own regulations and ensure that its reviewers consistently documented sex differences in NDAs (GAO, 2001).

To address some of these limitations, between 2002 and 2005, the FDA issued multiple recommendations for the inclusion and safety of pregnant women in clinical trials. In 2002, an FDA regulation on establishing pregnancy exposure registries provided guidance on monitoring the outcomes of pregnancies exposed to specific medical products with the goal of providing clinically relevant data to medical providers for treating patients who are pregnant (FDA, 2002). In a 2004 guidance, the FDA provided a basic framework for designing and conducting pharmacokinetic and pharmacodynamics studies in pregnant women, and provided instructions on how to assess the influence of pregnancy on pharmacokinetics and pharmacodynamics of medical products (FDA, 2004). A final draft guidance was released in 2018 titled "Pregnant Women: Scientific and Ethical Considerations for Inclusion in Clinical Trials" that supported an informed and balanced approach for gathering data on the use of medical products during pregnancy by encouraging judicious inclusion of pregnant women in trials and careful attention to fetal risk (FDA, 2018).

A push for diversity spurred Congress to pass Section 907 of the FDA Safety and Innovation Act of 2012 (FDASIA) (P.L. 112-144) that directed the FDA to investigate how well demographic subgroups were included in clinical trials and whether subgroup-specific safety and efficacy data were available (FDA, 2012). This law also required the FDA to provide Congress with an action plan that addressed improving the completeness and quality of data analyses on demographic subgroups. To fulfill these directives, the FDA drafted a report, *Collection, Analysis, and Availability of Demographic Subgroup Data for FDA-Approved Medical Products*, to address the extent to which demographic subgroups participated in clinical trials and whether the relevant subgroup analyses were performed in a manner consistent with FDA regulations (FDA, 2013). The FDA found variability across medical product types in the extent to which demographic data were analyzed. In some applications, subgroup analyses were limited by low sample size. Racial minority population subgroups were often underrepresented in trials. Communication of demographic information to the public also tended to vary for medical devices compared with drugs and biologics due to differences in the FDA regulatory frameworks.

In 2014, the FDA released *Action Plan to Enhance the Collection and Availability of Demographic Subgroup Data*, as necessitated by Section 907. This action plan outlined recommendations for the inclusion of demographic data in labeling and the public availability of these data. It also included new guidelines to encourage greater demographic subgroup inclusion in trials, plans to work with sponsors to improve information on demographic subgroups in NDAs and INDs, and intentions of strengthening FDA reviewer training by adding education in inclusion, analysis, and communication of clinical data (FDA, 2014b). In another 2014 guidance, the FDA outlined its expectation for sex-specific patient enrollment, analysis of the data, and reporting of the study information with the intention of improving the quality and consistency of data. Through this guidance, the FDA encouraged sponsors to investigate reasons for the lack of enrollment of women and suggested measures to correct this imbalance. For example, if women's participation dropped substantially after the initial trial screening, then the study criteria may have to be examined to reduce the unintentional exclusion of women. The guidance also provided recommendations to improve enrollment such as targeting investigation sites where women could be more easily recruited, considering alternative communication strategies for recruitment, and maintaining open enrollment for women until a target proportion has been achieved (FDA, 2014a).

Although several FDA regulations in the 1990s addressed the inclusion of women in trials, fewer regulations specifically targeted the inclusion of racial and ethnic minority population groups. When the Office of Women's Health sought to raise the issue of terminology for race in the 1990s, the FDA initially exempted itself from OMB definitions of race. The Office of Women's Health raised the issue again in 2004 when it drafted the first guidance around inclusivity on race that adopted the OMB definition of race and ethnicity for reporting trial populations. This draft was not finalized until 2016 (Wood, 2021).

The Office of Minority Health was only established in 2010 as part of the ACA to advise the FDA on reducing health disparities among racial and ethnic groups (DiPietro and Liu, 2016). In 2016, the FDA released the guidance titled "Collection of Race and Ethnicity Data in Clinical Trials: Guidance for Industry and Food and Drug Administration Staff" to provide instructions on the use of standardized terminology for demographic information (age, sex, gender, race, and ethnicity) based on OMB directives, to ensure that subgroup data was collected consistently (FDA, 2016). This was a very limited guidance that only discussed the terms used to describe "non-white" populations but did not explicitly encourage inclusion of these populations in trials. A guidance for medical devices was released in 2017, "Evaluation and Reporting of Age-, Race-, and Ethnicity-Specific Data in Medical Device Clinical Studies," in which the FDA provided recommendations for the evaluation and reporting of demographic-specific data in clinical studies (FDA, 2017). The guidance covered why diverse representation was important and identified potential barriers to enrollment, as well as provided recommendations to overcome those barriers.

Around this time, the FDA also began to require the use of Drug Trials Snapshots that provided information about the populations that participated in FDA-supported clinical trials, and highlighted whether there were any differences in benefits and/or side effects by sex, race, ethnicity, and age (FDA, 2020a). Although the use of Snapshots has marked progress in ensuring that demographic information is transparent and made available to the public, it has had limitations. The Snapshots only cover 2015 onward, and only provide information from Phase 3 studies or products that already have been approved (Wood, 2021). A final guidance was issued in 2020 titled "Enhancing the Diversity of Clinical Trial Populations – Eligibility Criteria, Enrollment Practices, and Trial Designs Guidance for Industry," specifically addressing the need to enhance diversity of clinical trial populations by modifying eligibility criteria, enrollment practices, and trial designs (FDA, 2020c). The guidance advises that drug sponsors have a "plan for inclusion of clinically relevant populations no later than the end of the Phase 2 meeting." Through this guidance, the FDA satisfied a mandate under Section 601 (a)(3) of the FDA Reauthorization Act of 2017 (FDARA) (P.L. 115-52) to broaden and develop eligibility criteria with no unnecessary exclusions for clinical trials, improve trial recruitment so that trial participants reflect the population that will use the drug, and apply these recommendations to clinical trials (FDA, 2020c).

This guidance also aimed at promoting enrollment of individuals with disabilities in clinical trials. For individuals such as older adults, disabled, or cognitively impaired individuals who need caregiver help or transportation, a requirement to make frequent visits to trial sites can be problematic and hinder participation. To make participation in clinical trials less burdensome, the FDA has proposed measures such as reducing the frequency of visits, considering whether visits could be replaced by telephone or virtual means, making participants aware of reimbursements for travel and lodging expenses for trial participation, and using mobile medical professionals to visit and evaluate participants or collect blood samples (Stephenson, 2020).

While historically the FDA has often used sex and gender synonymously, this guidance noted that the FDA recognizes that for some clinical trial participants, gender and sex may not be concordant. Still, it stated that discussion of this topic falls outside the scope of the guidance. Sexual minority populations remain largely overlooked by FDA policies (FDA, 2020d).

Enforcement, Incentives, and Accountability

Thus far, the FDA has undertaken various measures to improve diversity in clinical trials primarily via guidance documents and the use of Drug Trials Snapshots. Despite these efforts, certain demographic groups have remained underrepresented in many trials. The measures taken by the FDA have limitations that must be addressed so that populations participating in trials reflect the diversity of the population at large that will be using the drugs/medical products. One such limitation is the lack of enforcement of FDA guidance. Notably, most guidance documents contain the following disclaimer:

> FDA's guidance documents, including this guidance, do not establish legally enforceable responsibilities. Instead, guidances describe the Agency's current thinking on a topic and should be viewed only as recommendations, unless specific regulatory or statutory requirements are cited. The use of the word should in Agency guidances means that something is suggested or recommended, but not required.

The FDA encourages inclusivity but lacks the power to enforce recommendations made in published guidance documents. Although the FDA can make recommendations, it has limited capacity to enforce them. This dynamic may have developed in part because the FDA does not fund investigational drug trials, but rather assesses them, giving the agency less bargaining power compared with the NIH, which can provide funding. However, this is not an insurmountable barrier, especially since the 2000 law gave the FDA the power to put a clinical trial on hold if men or women were being excluded due to reproductive potential. Although the FDA does not appear to have ever put a trial on hold for this reason, such regulations can be leveraged to ensure that sponsors do everything in their power to improve enrollment practices and increase ease of enrollment so that clinical trials can be more inclusive. In addition, FDA also has authority within the IND and Investigational Device Exemption processes to provide feedback to sponsors of clinical trials. The IND process is needed for sponsors to be able to ship the investigational drug across state lines, and involves determining whether the product is reasonably safe for initial use in humans and that the pharmacology of the compound justifies its use commercially (FDA, 2021a). Given the safety concerns of not including a diverse population in drug testing, this process could be leveraged by the FDA to hold investigators accountable.

FDA progress has also been slow and tends to be reactive to congressional prodding. Developing a guidance is a long and resource-intensive process, and often there has been a lack of bandwidth to produce guidelines to improve inclusivity in trials. Such guidelines have been created in response to laws from Congress, but in the absence of such legal directives there has been limited planning to promote diversity. Changes have tended to arise not from leadership within the agency but from congressional action. The lack of internal initiative and leadership to independently create policies for inclusion has hampered progress toward diversity. In addition, too often if gender or race/ethnicity differences are not proven (e.g., difference in symptoms of a heart attack in women), then they have been treated as though absent (Wood, 2021). A more cautious approach could help to increase concern for the lack of diversity in clinical trials. Finally, the lack of standardization of submission of NDAs and INDs has been problematic. Since every sponsor tends to design trials differently and can label trial characteristics differently in the database (Wood, 2021), there has been no clear standard of submission. Thus, analyzing data and obtaining intersectional data from the FDA has been challenging. Setting guidelines for submission across trials of medical products and drugs could help standardize data collection, which could further aid in analysis. Although the FDA has made major progress in ensuring more inclusivity within clinical trials, it must continue to try to enforce its recommendations to ensure that trial populations reflect the diversity of populations at large.

In addition to enforcement and accountability measures, FDA could do more to incentivize industry to have more diverse clinical trials. This would encourage industry to fill an unmet need in drug development and there is precedent for these policy incentives. In 1983, Congress passed the Orphan Drug Act (ODA) to encourage industry to develop new drugs for rare diseases. The ODA provides tax credits to offset the cost of research and development for these rare diseases, waiving of Prescription Drug User fees, as well as a 7-year extension of market exclusivity for eligible products. This has undoubtedly led to an increase in the number of drugs for rare diseases, with over 800 drug indications approved between 1983 and 2019, compared to only 38 drugs prior to the passage of the ODA (Aitken et al., 2019). An analysis found that the ODA led to "a 69% increase in the annual flow of new clinical trials for drugs for 'traditional' long-established rare diseases" (Yin, 2008). Similar incentives could be developed to increase the diversity of clinical trials and clinical research and incentivize developing research infrastructure that includes communities from the outset of the research design.

However, these policies would need to be developed carefully to minimize costs for the public and patients, with an observation for lessons learned from the ODA. While the ODA has done a great deal for rare diseases, there are growing concerns about its costs. As the number of treatments for orphan drugs increases, there are increased concerns about the high costs of orphan drugs, their potential to threaten insurance premium levels, and their accessibility for many patients due to their high costs (Pearson et al., 2022). In addition, a 2018 GAO report found inconsistent and incomplete reviews in the process of designating medicines as orphan drugs, indicating that drug companies may be overusing the classification to maximize profits (GAO, 2018b).

The committee feels that some of these incentives could still be implemented with some guardrails in place that may minimize some of the downsides of the ODA. For example, a recent white paper analyzed some policy options to prevent some of these costs, including establishing a maximum revenue threshold to be eligible for ODA incentives, using sliding scale bonuses or refunds depending on outcomes for a particular drug, using volume-based contracts to purchase large volumes of drugs for rare conditions, and more (Pearson et al., 2022). These provisions could be considered and written into any potential new legislation to minimize the risk of harm while still providing incentives to private industry to diversify clinical trials and clinical research.

Centers for Disease Control and Prevention

The Centers for Disease Control and Prevention (CDC) sponsors a wide range of public health initiatives, including providing funding for local and state public health agencies. The research mission of the CDC is to support public health studies, and it often sponsors retrospective examinations of public health issues. Funding clinical trials makes up a small portion of the CDC's research budget, but the agency still regulates recruitment of diverse demographics.

CDC policies promoting inclusion of diverse populations in research have been largely catalyzed by federal laws. The 1993 NIH Revitalization Act did not govern the CDC, but the legislation spurred the CDC to create its own policies on inclusion of women and diverse races and ethnicities in research participation (Geller et al., 2011). In 1995, the CDC issued the "Policy on the Inclusion of Women and Racial and Ethnic Minorities in Externally Awarded Research," applying to extramural research activities (CDC, 1995). In 1996, the CDC released "Inclusion of Women and Racial and Ethnic Minorities in Research," which applied to intramural research (CDC, 1996). The ordinances stated that women and members of racial and ethnic minority population groups should be adequately represented in all CDC research involving human subjects, in the absence of a compelling reason for exclusion. They further stipulated that women of childbearing potential should not be routinely excluded from research without proper cause. The policies provided guidance on how these goals for inclusion could be met by investigators. For example, if all diverse groups could not be included in a single study, multiple studies could be conducted. The policies also stated that it was not necessary to provide the statistical power to test hypotheses in all groups separately, but that if differences between groups are plausible, this should be tested in the study design. Further, study proposals should include discussion of the inclusion or exclusion of minority groups (CDC, 1995, 1996).

The CDC policy on inclusion of women and ethnic and racial minority populations was closely followed by the more general policy in 1997, "CDC Procedures for Protection of Human Research Participants" (CDC, 1997). This new policy restated an abridged version of the 1996 ordinance, placing it into context of protections of other policies for the protection of human subjects.

The CDC revised the 1996 "Inclusion of Women and Racial and Ethnic Minorities in Research" policy in 2010. This policy united the past separate policies on extramural and intramural research into a single policy. The revised policy strengthened the call for representation in clinical research by stating that direct efforts should be made to actively recruit and enroll women and minority population groups in all funded research (CDC, 2010).

Protections for children in clinical research were codified in the 1997 "CDC Procedures for Protection of Human Research Participants." The 1997 guidelines deferred to local and state laws on medical consent for minors to determine whether it is appropriate for a minor to participate in a research study. However, the policy also stated that the minimum requirements for consent in a local jurisdiction do not necessarily authorize a minor's involvement in a research study. The policy gives latitude to the study's institutional review board (IRB) to determine the ethical parameters for a minor's involvement in research. The IRB should weigh risks to the children in the study with the benefits the research may provide to children as a group. The ordinance also stresses the importance that research poses a "minimal risk" to children, as adjudicated by the IRB. It outlines minimal risk research activities that are usually acceptable for children, such as urinalysis, venipuncture, electroencephalography, and allergy scratch tests (CDC, 1997).

The CDC issued an explicit policy on children in research in 2006 (updated in 2011) titled "Inclusion of Persons under the Age of 21 in Research." According to this policy, research proposals must include a rationale to include or exclude persons under 21 in intramural or extramural research—similar to the provisions in the "Inclusion of Women and Racial and Ethnic Minorities in Research" policy (CDC, 2011).

While the CDC's public health research and initiatives have received a plethora of academic attention, there has been less research on the clinical trial policies of the CDC. The CDC devotes a smaller proportion of its funding to clinical trials, but further attention to this issue may be useful.

Centers for Medicare & Medicaid Services

In addition to the CDC, other federal agencies that sponsor or regulate clinical trials include the Department of Defense, the Agency for Healthcare Research and Quality (AHRQ), and the Centers for Medicare & Medicaid Services (CMS) (HHS, 2017). The role that CMS plays in funding diverse clinical trials is especially important. While costs of routine care to participants in trials are usually covered by Medicare, many costs are borne by the participants, decreasing participation especially among financially disadvantaged patients (Medicare advantage, 2009). For participants receiving Medicaid benefits, there were historically no federal mandates for clinical trial coverage, making it prohibitively expensive for many Medicaid beneficiaries to participate in clinical trials (Winkfield et al., 2018).

This changed in 2000, when President Clinton issued an executive memorandum directing the secretary of HHS to cover the routine patient care costs associated with clinical trials, as well as costs due to any medical complications (The White House, 2000). The Health Care Financing Administration (the predecessor to CMS) responded to the executive order with a policy specifying that Medicare would cover "routine" costs that accompany clinical trial participation, including diagnostic tests, hospital charges, and provider fees, but excluding reimbursement for "items and services provided solely to satisfy data collection" (CMS, 2000). The exception is thought to have created major barriers that prevented community providers from participating in clinical trial enrollment, which in turn disproportionately affected racial and ethnic minority population groups. CMS updated its clinical trial policy in July 2007 with clarifications and some additional coverage items. However, the policy still excludes items and services for data collection (CMS, 2007).

In 2014, CMS released an updated guidance that allows CMS to determine coverage of an item or service only in the context of a clinical study. In its coverage with evidence development for transcatheter mitral valve repair, for example, CMS explicitly noted, "study protocol must explicitly discuss subpopulations affected by the treatment under investigation, particularly traditionally underrepresented groups in clinical studies, how the inclusion and exclusion criteria affect enrollment of these populations, and a plan for the retention and reporting of said populations on the trial" (CMS, 2014b). Such requirements by CMS have the potential to change the landscape of clinical trial representation because millions of Americans served by Medicare who have been traditionally underrepresented in clinical trials could now be recognized and actively recruited. Additionally, the $2.3 trillion omnibus spending and relief package passed by Congress in 2020 guaranteed, for the first time, routine costs for clinical trials for Medicaid recipients by 2022, expanding access for many low-income participants (Takvorian et al., 2021). Still, transportation, time away from work, and other ancillary costs remain barriers that will need to be addressed for participants despite their coverage status. In addition, the committee is not aware of any studies that specifically examine the impact that these reimbursement policies have had on accessibility for participation in clinical trials, the extent that they are utilized by participants, and any barriers that remain for participants accessing these coverage options.

Finally, while the updated version of this policy mandated that peer review of study protocols should assess "the adequacy of plans to include both genders, minorities, children, and special populations as appropriate for the scientific goals of the research," it omitted sexual and gender minority populations (NIH, 2019b).

CMS and the Center for Medicare and Medicaid Innovation also conduct demonstration projects, which are pragmatic clinical studies testing the effectiveness of different strategies and financial/reimbursement incentives for quality or outcome improvement. These studies involve data collection across sites, and achieving diverse enrollment has been a priority.

Agency for Healthcare Research and Quality

The AHRQ is an agency within HHS that has a mission to improve the safety and quality of America's health-care system. The Healthcare Research and Quality Act of 1999 (P.L. 106-129) established the Office of Priority Populations within AHRQ. Subsequently, the AHRQ has required that all AHRQ-supported research includes priority populations unless a compelling justification is provided against inclusion. Priority populations initially included women, children, and racial and ethnic minority population groups; populations with special healthcare needs (chronic illness, disabilities, and end-of-life care needs); and elderly, low-income, inner-city, and rural populations (AHRQ, 2021).

In 2021, President Biden signed Executive Order 13985, titled "Advancing Racial Equity and Support for Underserved Communities through the Federal Government" and defined underserved communities as individuals who have been denied "consistent and systematic fair, just, and impartial treatment." Specifically, it identified "Black, Latino, and Indigenous and Native American persons; Asian Americans and Pacific Islanders and other persons of color; members of religious minorities; lesbian, gay, bisexual, transgender, and queer (LGBTQ+) persons; persons with disabilities; persons who live in rural areas; and persons otherwise adversely affected by persistent poverty or inequality" as underserved communities (Executive Office of the President, 2021). Subsequently, the AHRQ updated its Policy on the Inclusion of Priority Populations in Research (NOT-HS-21-015) to expand its definition of priority populations to match those groups identified in Executive Order 13985 (AHRQ, 2021).

The Patient-Centered Outcomes Research Institute (PCORI) is a U.S.-based nonprofit institute created through the ACA (IRS, n.d.). Its funding comes from the Patient-Centered Outcomes Research Trust Fund (PCORTF) authorized by Congress in 2010 under the ACA, and reauthorized again in 2020 under the Further Consolidated Appropriations Act (P.L. 116-94). With PCORTF funding, in 2014, PCORI launched PCORnet, a national patient-centered clinical research network (Fleurence et al., 2014). PCORnet published "Diversity and Inclusion in PCORnet: Need and Recommendations" to set guiding principles for diversity and inclusion in PCORnet. The guidance called for inclusion and prioritizing underrepresented groups affected by the outcomes of research, including "people of color, rural/inner-city populations, pregnant and lactating women, gender and sexual minorities, individuals with disabilities, and other audiences commonly underrepresented in clinical research" (PCORI, n.d.). PCORI has also undertaken numerous measures recently to expand its work toward diversity, equity, and inclusion. The agency created an internal steering committee that is developing a comprehensive action plan to enhance diversity, equity, and inclusion. PCORI is also developing a data collection strategic framework with attention to diversity and inclusion.

The Common Rule

Many federal agencies are subject to the Common Rule (45 CFR 46), most recently revised in 2018 (HHS, 2017). The Common Rule is the blanket federal policy for the protection of human subjects that IRBs are expected to follow. The Common Rule requires that selection of research subjects be equitable, but it does not further specify inclusive recruitment of diverse subpopulations. Rather, the rule states that IRBs should be particularly cognizant of the special problems of research with subjects vulnerable to coercion or undue influence, such as children, prisoners, individuals with impaired decision-making capacity, or economically or educationally disadvantaged persons, and that additional safeguards should be included in studies to protect the rights and welfare of these subjects. The rule requires that IRBs should be composed of diverse individuals, and if an IRB regularly reviews research that involves a category of subjects that is vulnerable to coercion or undue influence, the IRB should consider including one or more members knowledgeable about and experienced in working with these categories of subjects. The Common Rule also specifies particular ethical regulations for children and pregnant women. One notable change in the 2018 revision was the removal of pregnant women from the "vulnerable" category of research subjects. This was in response to criticism that women were being unfairly excluded from research studies, to the detriment of designing treatments for women (NIH, 2019a). The revision aimed to increase the participation of women in research studies and to improve the recommendations for prescribing interventions for pregnant women (Hurley, 2017).

Enforcement and Accountability of CDC, CMS, AHRQ, and Common Rule

U.S. government agency policies on inclusion are enforced through different means. For example, CMS has an expectation that all supported clinical studies demonstrate adherence to inclusivity requirements, and that the agency would not anticipate approving a study that does not meet the requirements (CMS, 2014). The AHRQ peer review regulation requires that reviewers of grant and contract applications include their assessment of the proposed inclusion plan for priority populations in evaluating the overall scientific and technical merits of applications. Similarly, the CDC requires that grant reviewers abide by CDC policies on inclusion and diversity. Beyond grant review, accountability to inclusion policies also comes from a research applicant's IRB and oversight from HHS's Office for Human Research Protections (OHRP). The Common Rule grants latitude to IRBs to review research projects under the Common Rule's requirement that subject recruitment is equitable (HHS, 2017). However, the extent to which this is actually accomplished by IRBs in practice can depend upon the individual IRB's commitment to and interpretation of this objective. While OHRP does not directly oversee compliance of individual research studies, it does issue written assurances of compliance to research institutions, such as universities and academic medical centers, which allow investigators to perform research within the institutions. If an institution is noncompliant with the Common Rule, OHRP can revoke its assurance.

SPECIAL POPULATIONS

The Congress shall have the power to regulate commerce with foreign nations, and among the several states, and with the Indian tribes.
– Article 1, Section 8, United States Constitution, September 17, 1787

Including special populations in research is critical for addressing research questions. According to a 2018 review, "when special populations have been included into clinical trials, numerous age-dependent, community, cultural and genetic features have come to light." However, including these populations on clinical research requires consideration and use of best practices, such as building trust, conducting clinical trials that are relevant to special population, providing incentives and compensation, as well as offering options for participants to easily opt-out of the research (Winter et al., 2018).

Who are special populations? "It was not until the establishment of NIAAA [National Institute on Alcohol Abuse and Alcoholism] in the early 1970s, that using the concept of special populations as a categorizing strategy (both for funding and for the development of treatment programs) became prominent" (IOM, 1990). There is no standard definition of special populations; depending on where the term is used—in social work, education, medicine, criminal justice, or human services—the groups included will be defined by different terms.

Sovereign Nations

An unknown, legal, complex, and unique relationship exists between the United States and tribal nations. Hundreds of treaties, the Supreme Court, the President, Congress, executive orders, and laws have created a fundamental contract between 574 sovereign tribal nations and the United States (*Federal Register*, Vol. 86, No. 18), which recognize tribal nations as sovereign "domestic dependent nations under the protection" of the United States, and as sovereign nations, "exercise inherent powers over their members and territory" (Executive Order 13084, May 14, 1998). It is important to understand sovereignty and how it will play a role in research. In 2004, Kalt and Singer wrote *Myths and Realities of Tribal Sovereignty: The Law and Economics of Indian Self-Rule*, wherein the effects of tribal sovereignty reflected a three-decade resurgence in Indian country (Kalt and Singer, 2004). What happened 30 years earlier that engendered the resurgence? The passage of the Indian Self-Determination and Education Assistance Act of 1975 (P.L. 93-638) provides for maximum Indian participation in the government and education of the Indian people; provides for the full participation of Indian tribes in programs and services conducted by the federal government for Indians, and encourages the development of human resources of the Indian people; establishes a program of assistance to upgrade Indian education; supports the right of Indian citizens to control their own educational activities; and for other purposes. The act also provides tribal nations with the opportunity (other purposes) to manage health-care services in their local area, providing medical, dental, and behavioral health-care needs to improve the health and well-being of tribal members.

Indian Health Services

Specific challenges exist when trying to recruit American Indians related to both their physical locations and the structure of the Indian Health Service (IHS). Greater than 75 percent of American Indians live in urban areas, while only 1 percent of IHS funds are allocated to clinics in these areas; opportunities to interact with possible researchers is thus significantly limited (HHS, 2022a). Additionally, most studies require IHS and/or tribal approval, in addition to standard IRB approval, requiring additional care and consideration by investigators for navigating this system (Giuliano et al., 2000).

Millions of acres of tribal lands have been given up to the federal government; nearly all the land was acquired via treaty or agreement with tribal nations. In return, the federal government promised to provide health, education, and general welfare for reservation residents. These promises are known as "the United States trust responsibility to all Indians." It is the federal government's responsibility to protect tribal treaty rights, lands, assets, and resources, as well as carry out the mandates of federal law concerning tribal nations (112th Congress, 2nd Session). In April 2020, the National Tribal Budget Formulation Workgroup requested an "adequate level of funding" for the IHS Fiscal Year 2022 Budget. The President's Fiscal Year 2022 Budget includes an increase of $2.2 billion dollars in discretionary funding for IHS, or 36 percent above FY 2021, which is the largest single-year funding increase for IHS in decades (IHS, 2021). Preventable and treatable diseases, lack of basic health system infrastructure, and past failed policies all have contributed to avoidable health disparities, decreased life expectancies (see Table 3-1), and maintains impoverished health conditions for reservations and tribal members (NIHB, 2020). The IHS is the federal agency that oversees and provides health care to tribal communities through Indian tribes, tribal organizations, and urban Indian organizations.

Before COVID-19, the IHS was already so underfunded that expenditures per patient were just one-fourth of the amount spent in the Department of Veteran's Affairs health-care system and one-sixth of what is spent for Medicare. IHS facilities are, on average, understaffed by 25 percent (GAO, 2018a). Direct care, or medical and dental care that American Indians and Alaska Natives receive at an IHS or tribal medical facility, is covered through health benefits from the IHS. When a patient requires care that is not available at the IHS or tribal clinic, purchased referred care (PRC) funds are used, approval of which depends on several factors, including confirmation of tribal affiliation, medical priority, and funding availability. However, IHS remains severely underfunded, which contributes to its inability to meet its mission of raising the health status of American Indian/Alaska Native people.

During FY 2019, the Oklahoma City Area Indian Health Service (OCA IHS) was an example of severe underfunding. The per-person PRC funding level was $311.20. This led to OCA IHS having to operate at a Priority I service care level, also known as life-or-limb service (NIHB, 2020). To qualify for PRC care, a patient must be in peril of losing either their life or a limb. Additionally, there are 42 Urban Indian Organizations (UIOs) for health care in the United States providing care to the 78 percent of American Indians/Alaska Natives who are not living on a reservation either permanently or temporarily. These UIOs receive less than 1 percent of the IHS budget, which is currently underfunded at less than 50 percent of need (NCUIH, 2019). The reauthorization extension of the Indian Health Care Improvement Act (IHCIA) (P.L. 94-437) was passed in 2010 as part of the ACA. However, after 4 years, provisions of the act continued to be unfunded (NCAI, 2016). It is not difficult to understand why tribal members experience significant health disparities.

Due to the factors listed above and centuries of mistreatment, mistrust among American Indians and Alaska Natives has grown. To reach these communities, it

TABLE 3-1 2020 Life Expectancy at Birth

	Years	Women	Men
American Indians/Alaska Natives	78.4	81.1	75.8
Non-Hispanic whites	80.6	82.7	78.4

SOURCE: U.S. Department of Health and Human Services, Office of Minority Health, https://minorityhealth.hhs.gov/omh/browse.aspx?lvl=3&lvlid=62.

is critical that researchers involving tribal communities understand *tribal sovereignty*, as it is a fundamental root of a tribe.

> Indian nations pre-exist the United States and their sovereignty has been diminished, but not terminated. Tribal sovereignty is recognized and protected by the U.S. Constitution, legal precedent, and treaties, as well as applicable principles of human rights. (Kalt and Singer, 2004)

As sovereign nations tribes have legal rights and privileges not afforded to other groups, tribes can regulate research involving tribal members and the use and ownership of their data. What is or is not agreed upon with one tribe will not necessarily transfer to another tribe. And, there are no tribal- or federal-wide agreements for the inclusion of or data protections for tribal members who individually enroll in clinical studies not specific to tribes or tribal nations, such as the 19,806 American Indian/Alaska Native who participated in NIH-funded research in 2017. There is little information available to educate investigators about the concerns surrounding tribal participation and therefore little to guide them on how to approach inclusion of Indigenous individuals in clinical studies that are not specific to these groups and their Native nations (Kalt and Singer, 2004).

> What is important to remember, researchers and the scientific process will benefit from better understanding of participants' social, economic, and cultural contexts which can only be done by taking the time to leave the institution and going to where the participants live. (Vigil et al., 2021).

Therefore, regardless of the difficulty, it is critical that research, education, and outreach in Indian Country continue to be brought to the tribal nations.

Winter et al. (2018, p. 58), in the section entitled "History, Context and the Ephemeral Nature of Trust," describe the importance of understanding the history and context for special populations to anticipate behaviors and attitudes from research participants. This applies to any culture or population throughout history.

> To us, any part of ourselves is sacred. Scientists say it's just DNA. For an Indian, it is not just DNA, it's part of a person, it is sacred, with deep religious significance. It is part of the essence of a person.
> – Frank Charles Dukepoo (Pumatuhye Tsi Dukpuh), 1943–1999[2]

---

[2] Frank Charles Dukepoo was an acclaimed geneticist at Northern Arizona University and the first Hopi to earn a Ph.D.

4

Barriers to Representation of Underrepresented and Excluded Populations in Clinical Research

The analysis draws substantially from a research paper commissioned for this study, written by Drs. Farah Acher Kaiksow, M.D., M.P.P., and Jocelyn Carter, M.D., M.P.H. The full research paper can be found online at: nap.nationalacademies.org.

The processes and infrastructure of medical research have led to important advances in medical knowledge and therapies that have improved many lives; however, the existing system has also served to reduce participation by a diverse population. This chapter presents an overview of the range of factors, operating at multiple levels (participant and community characteristics, individual research studies, the institutions that conduct research, and the broader landscape agencies and policies that govern research), that serve as barriers to inclusion of underrepresented and excluded populations in clinical research.

INDIVIDUAL AND COMMUNITY FACTORS

Individual and community factors are often cited as reasons for lack of inclusion of underrepresented and excluded populations in clinical trials. The evidence suggests, however, that many of these concerns misrepresent barriers to participation in research or are surmountable with effort from research teams, funders, and policy makers. In addition to the barriers to inclusion that are often present in the life cycle of an individual study, a range of cultural, historical, and community-level factors influence feasibility and implementation of clinical research and directly influence study recruitment and retention.

Willingness to Participate

Overall, lack of willingness to participate is frequently given as the cause of poor representation of some populations in research. However, the evidence on this issue is clear: Asian, Black, Latinx Americans, and American Indian/Alaska Native individuals are no less likely than other groups, and in some cases are more likely, to participate in research if asked (Adeyemi et al., 2009; Arega et al., 2006; Bieniasz et al., 2003; Bishop et al., 2011; Byrd et al., 2011; Byrne et al., 2014; Ceballos et al., 2014; Evans et al., 2010; Gadegbeku et al., 2008; Garber et al., 2007; George et al., 2014; Guadagnolo et al., 2009; Hillyer et al., 2020; Kaplan et al., 2015; Langford et al., 2014; Manders et al., 2014; McElfish et al., 2018; Murphy and Thompson, 2009; Murphy et al., 2009; Priddy et al., 2006; Sanderson et al., 2013; Sprague et al., 2013; Thetford et al., 2021; Trant et al., 2020; Webb et al., 2010; Wendler et al., 2006).

A 2014 national review that included more than 4,500 Asian, Black, and Latinx Americans who were eligible for cancer trials found the same willingness to participate among all groups and equal enrollment rates (Langford et al., 2014). Study participation willingness was similar across racial/ethnic groups for studies focused on HIV, despite early narratives of stigma and discrimination related to the illness. Among Asian American, Black American, and white American college students in Atlanta, a 2006 study found no difference in willingness to participate in an HIV vaccine trial (Priddy et al., 2006). Results with older patients are equally convincing: among a population of 417 HIV-positive Black and Latinx people (60 percent male) in Chicago with an average age of 43, 95 percent would either agree to or consider participating in a study (Adeyemi et al., 2009). In this analysis, the strongest predictor of participation was simply being asked.

Rural populations are increasingly recognized as underserved, with underrepresented individuals from rural areas particularly at risk for poor health outcomes.[1] Enrollment of rural populations into clinical research is especially challenging given structural barriers including access to health care and transportation issues. Yet people living in rural areas do not appear to be any less willing to participate, based on a large study of 5,256 people in Arkansas and a smaller study of 533 people in Alabama, Florida, Georgia, Louisiana, Mississippi, and Puerto Rico (McElfish et al., 2018; Thetford et al., 2021). Among the respondents in the Arkansas study, greater than 45 percent said they would participate in research if asked, with another 22 percent being undecided; only 32 percent said they would not participate (McElfish et al., 2018). The smaller multistate study further analyzed the data by ethnicity and rurality and found that among

---

[1] The committee would like to note that it is unclear from the literature whether frontier populations are included in the research definition of "rural." Frontier areas are sparsely populated rural areas that are isolated from population centers and services, and there is no universally accepted definition of rural that ensures frontier populations are included in this demographic (Coburn et al., 2017). When this report was written, the committee could not find any literature specific to frontier populations' participation in clinical trials and clinical research.

Black and Latinx residents of both rural and urban areas, 75 percent were willing to participate in research, but greater than 90 percent had never been asked (Thetford et al., 2021).

Population-specific studies confirm what the more general studies cited above suggest: underrepresented populations are not necessarily underrepresented because they are unwilling to participate. Attitudes of 204 Black men about a variety of types of clinical research, including surveys, focus groups, clinical trials, and genetic studies, found that 74 percent endorsed a willingness to participate (Byrd et al., 2011). Regarding specific willingness to be randomized in a surgical versus nonoperative study of spinal disorders, Black Americans expressed equal willingness to be randomized as white Americans (Arega et al., 2006). The same results have been found in studies on Black individuals' participation in HIV treatment trials, studies on aging, and recruitment for clinical trials on kidney disease (Evans et al., 2010; Gadegbeku et al., 2008; Garber et al., 2007). In the HIV treatment study, like the more general 2014 study mentioned above, the major barrier to participation of HIV-positive Black people was having never been asked (Garber et al., 2007).

Data suggest that Latinx populations may in fact be more likely to participate than other populations. A 2014 study of women in Texas reported that Latinx women were 44 percent more likely than non-Latinx women to participate in a gynecologic malignancy clinical trial (Manders et al., 2014). In New York City, Latinx patients were more than twice as likely to say they would join a cancer clinical trial compared with non-Latinx patients (47.7 percent and 20.8 percent, respectively) (Hillyer et al., 2020). A qualitative study of 59 Latinx men and women at the Texas-Mexico border demonstrated significant enthusiasm on the part of this group to get involved in research (Ceballos et al., 2014). "If I had the opportunity to participate in something like this, I'd love to," said one respondent.

Although not as extensive, studies of American Indians echo those of Black and Latinx Americans. A study of American Indian college students found that, depending on the specifics of the trial, anywhere between 63 percent and 84 percent would probably or definitely agree to participate in a cancer clinical trial (Sprague et al., 2013). Only in cases where a significant amount of travel or risk of a confidentiality breach existed did willingness drop below 50 percent. In a separate study comparing American Indians with Asian, Black, Latinx, and white Americans, there was no difference between the groups in refusal to participate in a cancer clinical trial (Guadagnolo et al., 2009).

Although stated willingness is comparable across these underrepresented groups, it might differ from actual consent and participation rates; however, evidence suggests these are at least equal. A literature review, published in 2006, combined data from 20 studies that examined the consent rates of people of different races and ethnicities; 18 of these studies took place either entirely or mostly in the United States, while the remaining two studies took place in Europe, Australia, or New Zealand (Wendler et al., 2006). Combining data from these studies to create a cohort of more than 70,000 individuals, this analysis found that Black and Latinx people had the same consent rates as white people. For clinical intervention studies, Latinx individuals actually had statistically significantly higher consent rates, 55.9 percent compared with 41.8 percent, respectively. A more recent study of 1,126 postpartum women in Philadelphia found that consenting women were actually more likely to come from underrepresented groups compared with those who did not consent (Webb et al., 2010).

The inability to channel the willingness of underrepresented individuals to participate in research has implications beyond lack of engagement in specific trials. In almost all papers on predictors of willingness to participate in research, prior exposure to or participation in research is associated with increased likelihood for participation and a more positive attitude toward research (Behringer-Massera et al., 2019; Byrne et al., 2014; Sprague et al., 2013; Webb et al., 2019). In a study of more than 7,800 people in Florida, the positive influence of prior exposure on future participation was higher for Black respondents than for white respondents (Webb et al., 2019). Unfortunately, misunderstanding or lack of knowledge about the willingness of underrepresented populations to enroll in clinical research has created a pattern by which failure of researchers and/or clinicians to ask these groups to participate contributes to their lack of enrollment, which further decreases their chances of future involvement, and thus the cycle continues.

Trust

Any conversation about the low participation rates of underrepresented individuals in medical research must include the issue of distrust and/or mistrust of the health-care system. Whether caused by distrust (an individual's sense that their trust has been violated by a specific act, person, or institution) or mistrust (a less specific but no less legitimate feeling that a person or institution may not be acting in an individual's best interest) (Griffith et al., 2021), the legacy of both historical and contemporary abuses in medical research is an important factor driving the lack of engagement of underrepresented populations with both health care and research. This holds true across a range of underrepresented groups, including Asian American, Black American, Latinx American, and Mexican American (Adeyemi et al., 2009; Behringer-Massera et al., 2019; Bonevski et al., 2014; Braunstein et al., 2008; Buchbinder et al., 2004; Bussey-Jones et al., 2010; Byrd et al., 2011; Corbie-Smith et al., 2002; George et al., 2014; Hardie et al., 2011; Haynes-Maslow et al., 2014; Hoyo et al., 2003; Hughes et al., 2017; James et al., 2017; Lor and Bowers, 2018; Moreno-John et al., 2004; Murphy and Thompson, 2009; Murphy et al., 2009; Newman et al., 2006; Occa et al., 2018; Scharff et al., 2010; Smirnoff et al., 2018).

In qualitative studies with Black Americans, those who decline to participate or express lower willingness to participate frequently mention the offenses committed by the Tuskegee Syphilis Study as well as more recent personal stories of distrust as reasons for their declinations (Alsan and Eichmeyer, 2021; Behringer-Massera et al., 2019; Buchbinder et al., 2004; Byrd et al., 2011; Corbie-Smith et al., 2002; Scharff et al., 2010). The authors of a survey study about differences in willingness to participate in a cardiovascular drug trial suggest that in addition to not being asked, this type of distrust/mistrust can explain much of the participation gap between Black and white Americans (Braunstein et al., 2008). In a clinical trial exploring barriers and motivators to participation in clinical trials among 67 Black Americans, focus group themes included the perception that research would benefit white participants or the research institution more so than any underrepresented individuals enrolling in the study (BeLue et al., 2006). A study of 17 Black women at high risk for HIV found that, despite expressing favorable attitudes toward medical research in general, distrust was a commonly cited reason for not participating (Voytek et al., 2011). Similar studies exist regarding Black individuals' participation in blood/tissue donation for genetic studies (Bussey-Jones et al., 2010), psychiatry research (Murphy et al., 2009), and cancer research (Haynes-Maslow et al., 2014). These studies propose historical abuses as a major source of distrust among Black Americans and further assert that this distrust is a large factor in their unwillingness to enroll in medical research.

This issue of trust is of course not limited to Black Americans, and reasons for the distrust vary depending on the group or individual. In interviews with an older population of Hmong individuals, specific concerns arose about possible researcher misuse of information that might lead to loss of financial support from governmental agencies (Lor and Bowers, 2018). The Havasupai Tribe case regarding the misuse of genetic samples and lack of complete informed consent reinforced existing distrust of medical researchers and discouraged tribe members from participating in further genetic research (Garrison, 2013). In a study of 50 Filipino and Native Hawaiian/Pacific Islander people, major focus group themes included negative feelings about the purpose and intent of the research (Gollin et al., 2005). Research into Latinx Americans and Mexican Americans as well as Asian Americans of Filipino descent suggests that at least some of their distrust is rooted in fear for their own or a family member's immigration status (George et al., 2014; Hardie et al., 2011; Maxwell et al., 2005; Occa et al., 2018). Concerns about health insurance coverage have also been reported. A study of 88 Black Americans' attitudes toward genetic research identified fear of the loss of health insurance coverage because of targeted discrimination as a barrier to participation (Sadler et al., 2010). Populations who participate in illegal or culturally stigmatized behaviors, including intravenous drug users, people with substance use disorders, LGBTQIA+ individuals, and people who are HIV positive, also may not trust that their personal information will be kept private by research teams (Bonevski et al., 2014; Voytek et al., 2011).

Although the committee agrees that distrust and mistrust are certainly factors that influence the participation of historically underrepresented groups in clinical research, some studies have found that the distrust/mistrust is not associated with willingness to participate in medical research (Alhajji et al., 2020; Ford et al., 2008; Garber et al., 2007; Katz et al., 2008; Webb et al., 2019; Westergaard et al., 2014). A study of 5,139 Black individuals and 2,670 white individuals in Florida found that while Black respondents had mildly lower levels of trust in both researchers and research studies than white people, level of trust did not predict intent to participate for either group (Webb et al., 2019). Although mistrust is likely a factor, the studies that show it is not associated with willingness to participate in research may point to this not being an insurmountable problem and perhaps not the most important barrier. For example, a systematic review of 40 years of research on barriers to enrollment in cancer studies found that, although mistrust was the most commonly cited individual-level reason for not participating in research, the most common barriers overall were related to being offered the opportunity to participate.

Fundamentally, we may never know exactly how much historical and current discrimination and abuses influence underrepresented individuals' participation in clinical research. The research done in this area may be limited by participants' unwillingness to openly discuss trust issues with research teams that represent the very entities the participants distrust. Additionally, people with the highest levels of mistrust are unlikely to participate or to be represented in any research. Persistent and systemic efforts to delegitimize, underemphasize, or ignore the link between historical and contemporary occurrences of scientific misconduct/abuse and the mistrust of underrepresented populations toward research will certainly only continue to worsen current disparities in participation. Moreover, an inability or unwillingness of the research community to acknowledge and make efforts to address the roots of distrust/mistrust in underrepresented communities would stymie any movement toward increasing the trustworthiness of researchers in the view of underrepresented populations.

Social and Economic Factors

Although an individual's socioeconomic status is the result of a multitude of factors both within (individual level) and outside (structural level) their control, socioeconomic issues are discussed here at the individual level in an attempt to describe how these issues drive individual decision-making.

American women and underrepresented individuals make less money and are more likely to live below the federal poverty line compared with white men (DOL, 2020; KFF, 2019). Reduced economic resources can make elective participation in research a challenge. Jobs with fewer options for earned time, sick days, vacation days, and remote work may make participation in research impossible. Individuals with lower incomes are also frequently responsible for caring for children, elderly family members, and sometimes both at the same time, while also working outside the home (Indorewalla et al., 2021). Under these circumstances, even those individuals who do have the time to participate may not see the value in altering their regular routines, as this may pose much higher opportunity costs, including the loss of potential wages, than for those with more resources (Brown et al., 2000; Olin et al., 2002; Quiñones et al., 2020). One study on HIV/AIDS research in Black men with a history of drug use found that, despite their willingness, eligible participants were often not able to participate due to competing priorities related to work and family (Slomka et al., 2008). In focus groups and in-depth interviews with Asian American women assessing perceived barriers to participation in cervical cancer prevention research, reasons for nonparticipation included lack of time and inconvenience (Giarelli et al., 2011). Time conflicts and childcare responsibilities also emerged as barriers to participation in research for a study examining the perspectives of Black- and Latinx-immigrant participants (Calderon et al., 2006). A systematic review of barriers to study retention found that the most commonly reported barrier was competing priorities related to participants' socially disadvantaged status (Bonevski et al., 2014).

Perhaps most important are issues of opportunity costs, which include the loss of any potential gains that participants might be able to make if they choose to participate in research rather than the other potential activities. Whether it be a one-time 10-minute survey or a years-long clinical trial, study participation requires time away from work, family, and other commitments. Given this, household financial position plays an outsized role in who gets included in clinical research. Worldwide, nearly 50 percent of the people who participate in clinical trials are considered "high income," despite representing only 16 percent of the total population; conversely, the "lower middle class" makes up 38 percent of the population and 13.5 percent of the people who participate in clinical trials (Gilmore-Bykovskyi et al., 2021). A prospective study of cancer trials within the United States confirms that this global pattern holds true in the United States, even after accounting for factors such as age, race, and education (Unger et al., 2016).

Several studies have examined the importance of educational background or highest level of grade completion in research participation, many of which support the notion that educational status is more relevant than income level for the participation of Black Americans in research (Alhajji et al., 2020; Byrd et al., 2011). In a study of perceptions influencing research enrollment among low-income Black, Latinx, and white residents of New York City, respondents who had less than a high school education were more likely to have increased feelings of exploitation associated with research participation (Smirnoff et al., 2018). However, other studies have found no specific association with participants' highest level of education and willingness to participate in research (Kaplan et al., 2015).

Health literacy of patients and potential participants has been cited as a contributor to low participation in research, and low health literacy and numeracy skills are independently associated with less interest in research participation (Kripalani et al., 2019; Protheroe et al., 2009). However, as described earlier, although there is varied understanding of medical and scientific topics among individuals, it is possible to engage across a spectrum of participants, if appropriate efforts are made. A study of Asian, Black, Latinx, and white men with prostate cancer in California found no difference between people with low health literacy compared to those with medium or high levels of health literacy regarding their willingness to participate in clinical trials (Kaplan et al., 2015). The same study used a questionnaire to assess general knowledge of clinical trials and, again, found no difference in willingness to participate based on the respondent's understanding of research.

Challenges related to the frequent residential moves and lack of landline telephone access are also often cited as a primary reason for low enrollment and low retention of underrepresented populations (Bonevski et al., 2014; Otado et al., 2015). Reliable telephone access is a significant barrier for those living at or near the poverty line and has been associated with limited insurance coverage, healthcare access, and health behaviors (Bonevski et al., 2014). Perceptions of neighborhood safety have also been reported as reasons for reduced research participation of underrepresented individuals (Ceballos et al., 2014; Ejiogu et al., 2011).

INDIVIDUAL RESEARCH STUDIES

The factors and problems that lead to the limited enrollment of underrepresented and excluded populations in clinical trials and research begin with and follow the life cycle of a project. While a substantial body of literature describes individual and community characteristics (e.g., childcare needs or limited public transportation) that may prohibit research enrollment, these issues remain unaddressed well after the study is designed, funded, and under way. Understanding and resolving underrepresentation in research requires careful examination of the research process itself. At the level of an individual research study, there are problems and factors that prevent the inclusion of underrepresented and excluded populations in clinical research at almost every stage in the process, including

- the development of research questions;
- the composition, training, and attitudes of the research team;
- research site selection;
- participant selection, including sampling and recruitment methods and inclusion and exclusion criteria;
- study protocols, including informed consent processes and remuneration; and
- development of multilingual recruitment and consent documents.

The authors of a systematic review of 40 years of cancer treatment or prevention trials summed up the issue well, writing that "because opportunity barriers largely reflect protocol design as well as the process of study implementation, investigators play a major role in determining the extent to which trials are accessible to underrepresented groups" (Ford et al., 2008). In the sections below, the committee describes the many ways these problems and factors manifest throughout the course of the life cycle of an individual research study.

Research Questions: Drivers to Motivate Inclusion

Research questions are often driven by funding priorities and scientists' interests and expertise, which constrains the range of questions that are asked and answered. Laypersons (patients, community members) are rarely a part of the process of developing and refining a research question, even when they are representative of the population the research team proposes to engage or help. Engaging patients and community members can take a variety of forms, from advisory boards to pilot testers (screening forms, scientific measures, intervention components) to true collaboration on design, implementation, and analysis (with shared funding and ownership of data).

Patient and Community Engagement

Engaging patients, community members, or other stakeholders in research has been identified as a useful strategy for enhancing participation of underrepresented groups in the research process and, ultimately, reducing health inequalities and improving population health outcomes (Nguyen et al., 2021). For example, activism by the HIV/AIDS community led to the first federally funded community advisory boards (CABs) and galvanized the siloed research establishment (Karris et al., 2020). CAB recommendations led to trials of combination therapies instead of one or two drugs at a time, the creation of a participant's bill of rights, a robust informed consent process, and early vaccine trials (Strauss et al., 2001). The active involvement of CABs helped establish national research priorities, including emphasizing the needs of underserved groups such as women, and was considered critical to the overall quality of AIDS/HIV research (Karris et al., 2021; Strauss et al., 2001).

Patient engagement in research refers to patients or caregivers serving as partners or leaders in the research process, resulting in study decision-making that incorporates the experiences, expertise, and values of these stakeholders (Harrington et al., 2020). Better understanding of patient- and community-level concerns about research and their needs for participation in clinical trials can lead to more effective outreach tailored to specific individuals and populations and improved patient experience in clinical trials through less arduous screening, more responsiveness to inquiries from potential participants, and more attention to participants' needs (Forsythe et al., 2019; Smith et al., 2015). Systematic reviews indicate that patient engagement in research enhances study enrollment rates and increases participant retention (Crocker et al., 2018; Domecq et al., 2014). These patient engagement relationships may be very individualized and personal between a patient and clinician or care setting, unlike community engagement, which is likely built between an academic team and/or clinician that works with more than one individual (Kimminau et al., 2018). Community engagement in research involves inclusive participation of people affiliated by geography, sociodemographic characteristics, or shared interests (Wallerstein et al., 2018). Community-based participatory research, a form of community-engaged scholarship that emphasizes rigorous partnered processes and focuses on community priorities, has been associated with significantly higher recruitment and retention of minority participants in research (Las Nueces et al., 2012; Yancey et al., 2006) and better behavioral and clinical outcomes (O'Mara-Eves et al., 2013). In one systematic review of clinical research studies, patient and community involvement in designing recruitment and retention strategies, developing patient-facing information, helping to identify potential participants, or providing feedback on poor recruitment rates was associated with higher odds of a patient enrolling in a clinical trial (Crocker et al., 2018). An exploratory finding in these analyses was that the effect size was significantly higher when there was substantial involvement of patients or caregivers with lived experience of the condition being studied.

Among the commonly cited barriers to conducting patient- and community-engaged research are defining the community or patient partners for collaborations, capturing and addressing diverse viewpoints and perspectives, time and budget restrictions, and lack of researcher training in patient and community engagement strategies (Domecq et al., 2014; Levitan et al., 2018; Nguyen et al., 2021). Patient- and community-engaged research have been found to be feasible in many settings with careful planning, adequate training, and appropriate funding for the collaboration (Crocker et al., 2018; Domecq et al., 2014). Cost concerns may be mitigated by consideration of the financial benefits of these engagement approaches. An analysis of the financial value of patient engagement found that engagement activities (such as patient advisory panels or patient reviews of the protocol) can reduce the need for protocol amendments and their associated delays and costs, increase enrollment, and reduce study dropouts (Levitan et al., 2018). For a generic oncology new molecular entity, the study estimates that patient engagement activities that avoid one protocol amendment and improve enrollment, adherence, and retention is the equivalent of accelerating pre-phase 2 product launch by more than 2.5 years (and by 1.5 years for pre-phase 3).

It has been more than 50 years since the participatory research paradigm gained traction in the social and health sciences (Wallerstein and Duran, 2006). In the intervening years, most academic institutions, health organizations, and funders have recognized the need for and required, at least to some extent, community engagement in research. However, the pressures of academia and scientific research often preclude meaningful engagement of communities, which can be a slow and challenging process. Among these pressures are discipline-specific and institutional definitions of rigor and productivity, funders' focal interests, timelines, and financial investments in research projects, and the researcher's own scientific and professional interests (Cornwall and Jewkes, 1995). For example, time for building partnerships for a single study is typically not possible given the constrains in budgets and the limited timelines of grants. This makes it challenging to do partnership building, data collection, implementation of the study, analysis, and community dissemination under one contract grant. Nevertheless, many institutions and departments have CABs, community representatives on the institutional review board, and/or institutes that focus on community-academic partnerships. Federal funding agencies, like the National Institutes of Health (NIH), have similar mechanisms for public engagement in the research enterprise (Agnew, 1998). The proliferation of these features of the academic landscape is an acknowledgment of the importance of community engagement, but true engagement and empowerment requires an approach focused on co-learning and generating knowledge, rather than perfunctory stops along the research trajectory.

Research Team

Research shows that health outcomes are improved when a patient and physician are of the same race. Alsan et al. (2019) found that Black doctors could reduce the Black-white male gap in cardiovascular deaths by 19 percent. Additional research has shown that hiring diverse staff and providing proper training for clinical staff are important facilitators for improved recruitment and retention of diverse clinical trials participants (Butler et al., 2013; Quinn et al., 2012). This is also one of the main facilitators to successful recruitment and retention of underrepresented populations in clinical research from the interviews discussed in Chapter 5. However, there is little research on, and therefore little evidence to suggest, that concordance between participants and the clinical workforce would increase participation in clinical research. Some studies have shown that diverse staff do not play a key role in participation in clinical research. For example, in one study of adults living with HIV, only 12 percent of respondents felt that having a research staff of the same race was important (Adeyemi et al., 2009). In another study of Black Americans who either elected or declined to participate in a study on kidney disease found that neither the gender nor the ethnicity of the recruiter had any influence on likelihood of enrollment (Gadegbeku et al., 2008). Similarly, a study promoting group management of heart failure among Black individuals found that most participants did not request a Black group leader (Rucker-Whitaker et al., 2006). However, some Black participants asked that the people helping to manage their diets provide culturally relevant suggestions. This may suggest that it is most important for staff to be able to give advice and relate to populations represented in clinical research, rather than being from the same racial or ethnic group as them. For example, respondents to a survey of Hmong-speaking people said that speaking the same language was less important for participation than having a trusting relationship with researchers who were known and had created relationships within their communities (Lor and Bowers, 2018). This was also a finding from the interviews on facilitators to successful recruitment discussed in Chapter 5, which found that cultural and linguistic congruence with the target population was not enough and that gaining engagement and community buy-in for the study goals and desired outcomes was equally important.

Engaging with participants and building relationships requires genuine respect for individuals and their communities. The clinical psychologist Carl Rogers advocated for having an attitude that is "non-evaluative, nonjudgmental, without criticism, ridicule, depreciation, or reservations" for the patient (Patterson, 1985). This does not mean that physicians should change their values, but should not impose their own values and demand change from participants simply because they are the medical expert. In an editorial, Frosch and Tai-Seale (2014) suggest that "instead of lecturing (whether mentally or verbally) non-adherent patients, physicians can humbly inquire and ask the patient to reveal the reasons behind their behaviors, from which the physician can learn the barriers and identify potential levers for change."

Investigator Biases

Despite a demonstrated willingness to participate in research, underrepresented populations are often not asked by researchers to participate in clinical studies (Adeyemi et al., 2009; Byrd et al., 2011; George et al., 2014; Katz et al., 2006; Murphy and Thompson, 2009; Webb et al., 2019). A contributing factor appears to be attitudes of the research staff and health-care providers who are responsible for recruitment. There is evidence that, while acknowledging the importance of diversity in an abstract way, many principal investigators may not see diversity as an important factor in their own work. A 2020 study of 313 researchers at a large research university found that while 87 percent of respondents believed that diversity was very or extremely important, only 38 percent reported that it was a priority in their own research programs (Passmore et al., 2020).

Principal investigators and study staff also bring their own biases to the research enterprise. Their perceptions about a potential participant's reliability, health literacy, language skills, and social support, among other factors, all play into whether the potential participant will be offered information on enrollment (Joseph and Dohan, 2009). In one study, 92 percent of HIV/AIDS researchers felt that individuals with substance use disorders would need more support during trial participation than so-called traditional participants; 50–60 percent of these researchers believed that Black and Latinx individuals, as well as women, would also need additional support (King et al., 2007). In the same study, these researchers also had biases about their perceptions of the willingness of different groups to enroll in studies: 77 percent felt that white men were generally highly interested compared with 33 percent for white women, 20 percent for Black men, 16 percent for Black women, 13 percent for Latinx men, and 11 percent for Latinx women—these numbers stand in contrast to the results of studies on willingness to participate, which show a high degree of willingness to participate in research among women and underrepresented minority populations (Adeyemi et al., 2009; Byrd et al., 2011; George et al., 2014; Katz et al., 2006; Murphy et al., 2009; Webb et al., 2019).

Investigator and staff biases may influence the amount of time and effort they expend recruiting participants from underrepresented populations. In one study, oncologists used far fewer words and spent significantly less time with Black patients than with white patients, in both the clinical care visit and discussion of clinical trial enrollment (Eggly et al., 2015). Additionally, discussion of clinical trials was less robust for Black patients, with more emphasis on voluntary participation and less focus on the purposes and risks of participation. A different survey of Black cancer patients found that only one-third of eligible patients reported being given written information on possible clinical trials (Brown et al., 2013). Among persons living with HIV/AIDS in the United States, Latinx respondents were less likely to know about research opportunities compared with both white and Black respondents, and Latinx and Black patients were less likely to be notified about possible enrollment by any member of any clinical or research team (Castillo-Mancilla et al., 2014).

Site Selection

Several studies have determined that the distance to health care and clinical research from a patient's home, or home community, is also a factor that prevents participation. Most clinical research takes place at or near large academic centers that are less frequently used by some underrepresented populations compared with community health settings. The greater the distance between home communities and where patients are required to present for initial involvement, study visits, or exit interviews, the less likely they are to participate (Coakley et al., 2012; Sprague et al., 2013; Unger et al., 2016). Given the issue of distance, challenges with transportation have also been identified among the most common reasons for not participating in research studies (Brown et al., 2000). This relationship has been specifically established for Native Hawaiian and Pacific Islander populations, where individuals may be more likely to be living in remote areas and in under-resourced settings away from where research usually takes place (Giuliano et al., 2000). Conversely, a qualitative study assessing the effectiveness of offering transportation via a research van that would pick up participants in their home communities and then drive them to the research study site found that participants were highly satisfied with the convenience that transportation offered (Alcaraz et al., 2011). Research activities that do not offer transportation thus do so at the risk of excluding those without access.

Participant Selection

Sampling and Recruitment Methods

Another factor that prevents recruitment of racial and ethnic minority population groups in research is the way that recruitment is typically performed. Sampling methods may decrease the chances of diverse enrollment. Often, random sampling methods simply do not result in large enough study populations to capture the needed diversity. Random sampling can miss people who may want to remain hidden for a myriad of reasons (e.g., fear of discrimination, prosecution) such as LGBTQIA+ individuals or people with substance use disorders (Bonevski et al., 2014). Different recruitment methods have been shown to work for different populations. Mass media, including television, radio, and newspaper ads, may work well for one group, while word-of-mouth is much more suitable for another (Bistricky et al., 2010; Coronado et al., 2012; King et al., 2011).

Inclusion and Exclusion Criteria

Another element of the existing research structure that serves as a factor or problem preventing participation for diverse research populations is the development and application of inclusion and exclusion criteria that restrict or undermine the inclusion of underrepresented and excluded populations. Eligibility criteria must be carefully designed and intentionally applied to address the question being evaluated and achieve accurate and meaningful results, yet these restrictions often lead to the unintentional and systematic exclusion of certain groups (Langford et al., 2014; McKee et al., 2013; Quiñones et al., 2020). For instance, asthma researchers trying to assess differences in bronchodilator response found that potential participants from underrepresented groups were more likely to have inadequate responses to the methacholine challenge, one of the inclusion criteria; however, the methacholine challenge cut-point may have lacked sensitivity for underrepresented populations given previously reported differences in methacholine responsiveness among different racial/ethnic groups (Hardie et al., 2010). Similarly, a lack of preexisting or baseline data may result in unintentional exclusion of underrepresented participants. Initial chart review to determine eligibility for a study on COPD (chronic obstructive pulmonary disease), for example, unintentionally missed patients without baseline spirometry data, despite the designers' intentions to minimize exclusion criteria and maximize enrollment of underrepresented populations (Huang et al., 2019).

Lack of access to adequate health care is more common among underrepresented populations and can lead to delayed diagnoses and a more advanced form of disease, which can make individuals ineligible for study enrollment (Giuliano et al., 1998; Ward et al., 2004). Review of cancer trial recruitment among a medically underserved population that included American Indians found that restrictive inclusion criteria was one of the most common reasons for lack of enrollment

(Guadagnolo et al., 2009). Among the 88 potential American Indian participants, advanced stage/poor performance was the most commonly cited reason for non-enrollment (27 percent).

An analysis of the exclusion criteria for a study on smoking cessation found that Black and Latinx patients were more frequently excluded than white patients (Hooper et al., 2019). Additionally, in this analysis, white patients were usually excluded for a single reason, such as serious mental illness, difficulty with attendance, or medical conditions, whereas Black patients were more than twice as likely to be excluded for three or more reasons, such as smoking status, barriers to attendance, lack of motivation, or other health contraindications. Another report, also on eligibility for a smoking cessation study, found that despite being nearly twice as likely as white contacts to complete initial telephone screening, Black contacts were less likely to be eligible for enrollment (King et al., 2011). This difference persisted even when controlling for demographic factors such as education, gender, and income level. These analyses illustrate how the structure of current inclusion and exclusion criteria, intentionally or not, reduces opportunities for underrepresented individuals to participate in research.

Research Processes

Researchers are often not trained or skilled in explaining research methodologies or the potential positive impacts of research outcomes in ways that actively engage ethnically underrepresented populations (Bonevski et al., 2014; Hughes et al., 2017). Studies examining publicity and advertising for recruitment into research studies have identified a general failure to message the positive implications of research outcomes. Yet, there is evidence that this problem can be solved or mitigated. In a qualitative study exploring reasons for consent or refusal to participate in a comparative effectiveness study, researchers found that further explaining how a comparative effectiveness study works—for example, emphasizing that it does not test new medications—increased respondent's positive views of the study (Behringer-Massera et al., 2019). A group of researchers in Baltimore, Maryland, were able to successfully recruit a diverse cohort of more than 3,700 participants into a 20-year longitudinal study on aging in part by focusing on the direct benefits to the enrollees of their participation (Ejiogu et al., 2011). Similar studies have shown increased interest in research when people believe the research might provide personal, familial, or societal benefits (Boise et al., 2017; Gadegbeku et al., 2008).

Consent Processes

Finally, the length and complexity of the research process, especially the consent process and consent forms, has also been reported as a factor that prevents enrollment for underrepresented populations (Durant et al., 2014; Nipp, Hong and Packett, 2019; Hamel et al., 2016; Langford et al., 2014). In one study of people living with HIV that included predominantly Black and Latinx individuals, 19 percent cited the consent form being too hard to understand as a reason why they did not participate (Adeyemi et al., 2009). However, this barrier is not unique to underrepresented populations, as overly complicated consent forms are a barrier for all groups to participate in research (Kass et al., 2011; Sauceda et al., 2021).

The long time frame of most research projects may also reduce willingness to participate or remain enrolled in a clinical trial. A model created using data from potential cancer research participants found that the longer the time between a potential participant's consent to first contact by a study team member predicted probability of attrition; this effect was higher among racially underrepresented people compared with white individuals (Azfar-e-Alam et al., 2008). Current consent processes and consent forms are linguistically and culturally inappropriate for many underrepresented groups (Sauceda et al., 2021).

Additionally, researchers need to provide more appropriate recruitment materials, tailored to the language and literacy needs of potential research participants. The lack of suitable study materials in their respective languages has been shown to reduce participation of Asian, Creole, Hmong, Latinx, and American Indians, as well as Native Hawaiian/Pacific Islanders (Byrne et al., 2014; Calderon et al., 2006; Giarelli et al., 2011; Giuliano et al., 2000; Huang et al., 2013; Lawrence, 2000; Lor and Bowers, 2018; Nguyen et al., 2005; Occa et al., 2018; Tu et al., 2005). Even when language-specific materials are available, the quality and integrity of those materials may not be high. For many languages, verbatim translations are unlikely to capture the true meaning of the materials without incorporating commonly used idioms and culturally appropriate phrasing. This may especially be true for Spanish-speaking groups, as there are significant differences among the languages spoken in different Spanish-speaking countries (Occa et al., 2018). Translations that do not reflect the appropriate dialect or accepted verbal usage patterns can further discourage targeted populations from enrolling in a study. Fundamentally, it is critical that linguistic and literacy needs of diverse research participants are met.

All of the above-mentioned challenges—study design, outreach methods, choice of incentives, and research processes—are exacerbated by time and financial restrictions placed on researchers. Prioritizing speed, combined with a historically uninformed approach to minority recruitment, has led to a system in which research trials do not adequately prioritize enrollment of underrepresented populations.

Health-Care Access and Strong Primary Care

Closely related to socioeconomic status is access to health care. Lack of or limited health-care access is a root cause of inequitable health care throughout the United States. In a recent study of individuals just before and after age 65 (age of Medicare eligibility), Wallace et al. (2021) found that those eligible for

Medicare showed a marked reduction in racial and ethnic disparities of insurance coverage, access to care, and self-reported health. Besides obvious health-care consequences, this inequity also has implications for research. Patients who are not actively engaged with the health-care system will have limited opportunity for enrollment in studies.

Strong, trusting relationships with primary care providers (PCPs) have been noted to have significant impacts on research engagement (Adeyemi et al., 2009; Buchbinder et al., 2004; Friedman et al., 2020; Gadegbeku et al., 2008; Trantham et al., 2015). One study performed in five geographically diverse health-care centers (New York City; Baltimore, Maryland; Birmingham, Alabama; Iowa City, Iowa; and Boston, Massachusetts) found that the positive endorsement of a PCP led to increased likelihood of participation, while a negative attitude almost always led to a refusal to enroll (Buchbinder et al., 2004). In another study, simply having a PCP was the strongest predictor of clinical trial follow-up among a population of predominantly ethnically underrepresented individuals; socioeconomic status was not significantly associated with follow-up (Friedman et al., 2020). A North Carolina study on the involvement of Black male cancer survivors in research found that these patients and their families expressed significant trust in their physicians and would be open to enrollment in a research study if their physician suggested it (Trantham et al., 2015). Conversely, patients who are reluctant to visit their PCPs are more likely to be nonparticipants in medical research (Gadegbeku et al., 2008). This pattern also holds for other members of the health-care team, such as nurses, with patients reporting that they would not participate in a trial if their nurse does not recommend it (Adeyemi et al., 2009).

As described earlier in this chapter, several studies have also determined that the distance to health care and clinical research from patient home or home communities is also a problem, since most clinical research takes place at or near large academic centers that are less frequently used by underrepresented populations compared with community health settings. The greater the distance between home communities and where patients are required to present for initial involvement, study visits, or exit interviews, the less likely they are to participate (Coakley et al., 2012; Sprague et al., 2013; Unger et al., 2016).

LANDSCAPE FOR RESEARCH—COMMUNITY AND POLICY FACTORS THAT INFLUENCE THE REPRESENTATIVENESS OF CLINICAL TRIALS AND RESEARCH

Diversity interdigitates with each stage of the clinical trial and clinical research process. In the ideation stage, some questions might not be asked if there is not diversity among principal investigators and faculty driving the research questions, as described above. In addition, diversity of studies will also be affected by where the site is chosen for recruitment, and how it occurs.

The larger research enterprise and environment required to support diverse research studies present additional factors and problems that prevent the inclusion of a diverse research population.

Research Infrastructure to Facilitate Diversity in Clinical Trials and Clinical Research

Academic Medical Centers

As of 2019, academic medical centers comprise the nation's 154 accredited medical schools and more than 400 major teaching hospitals and health systems. These institutions conduct 55 percent of the extramural medical research supported by the NIH and operate 98 percent of the nation's 41 comprehensive cancer centers (Fisher, 2019). As such, these centers have substantial influence on the clinical trial enterprise.

Nevertheless, the traditional academic medical center structure creates substantial barriers to adequately consider diversity, equity, and inclusion in clinical trials and research. Sustainably and meaningfully engaging underrepresented and underserved populations often does not align with the traditional paradigm of promotion and tenure. Traditional academic centers mostly value teaching, research, and service, and a researcher's success is mostly judged by their productivity in publishing research and obtaining grant funding. Applying principles of community-based participatory research (see the Patient and Community Engagement section, above) and recruiting diverse population groups into clinical trials and research is time-consuming and requires investments to build and sustain trust; these investments are often only minimally considered in promotion and tenure decisions. This scenario often creates little incentive for early-stage investigators to invest time and resources to build community relationships. Moreover, academic institutions often provide inadequate institutional resources for researchers to engage communities, especially beyond the lifespan of a single research project.

Additionally, recruitment and retention of diverse faculty and staff are challenges in many academic medical centers. Research shows that women, particularly those from racial and ethnic minority groups, as well as men from racial and ethnic minority groups, are underrepresented in medical faculty. In an analysis conducted by the Association of American Medical Colleges, only 3.6 percent of medical school faculty are Black, 3.2 percent are Hispanic, and only 0.2 percent are American Indian or Alaska Native (AAMC, 2020). Further, although overall, women are about at parity with men at the medical school level, women leave the profession as they move throughout the career pipeline. For example, in 2019, women composed 48 percent of medical school graduates but only 41 percent of the full-time faculty. Further, women made up only 37 percent of associate professors and 25 percent of full professors in academic medicine. The percentage of women chairs and deans is even lower, at 18 percent for both positions (AAMC, 2020). In addition, racial and ethnic subgroups of women face a double bind in medicine and are even more underrepresented at higher academic ranks. Although racial and ethnic subgroups of women represent 18 percent of the U.S. population, only 3.2 percent of full professors in medicine are women from racial and ethnic subgroups (Carapinha et al., 2017; NASEM, 2020a).

Although academic medical centers have a long-standing history of community service, many underrepresented and underserved communities lack trust in these institutions. Academic researchers have been referred to as "in-and-out" researchers or "parachute" researchers (Stefanoudis et al., 2021), where one takes and does not give back to the community that has enabled the research success. Wilkins and Alberti (2019) argue that to address health inequities and truly engage with communities to address their research needs, academic health centers will need "commitments from institutional leaders, infrastructure to support engagement, and changes in policies to fuel innovative partnerships, facilitate community partner integration, and reward community-engaged scholarship." There is a growing need for academic medical centers to shift from community service to an "enterprise-wide approach" to community engagement in order to advance their missions of clinical care, education, and research. This shift in approach is critical to understanding, examining, and addressing the social determinants of health and structural barriers relevant to underrepresented and excluded communities.

Engagement opportunities across academic health centers and their benefits are described in Table 4-1.

Community Health Centers

Other types of health services organizations, including community health centers and rural health centers typically provide care to diverse population groups and represent an untapped resource for clinical trial and research recruitment. Federally qualified health centers (FQHCs) are grantees of the Health Resources and Services Administration, under Section 330 of the U.S. Public Health Service Act (P.L. 78-410), and include migrant health centers, health care for the homeless health centers, and public housing primary care centers. FQHC Look-Alikes meet all the requirements of health centers and reap most of the benefits of health center status, but do not receive a federal grant. Rural health centers can be public, nonprofit, or for-profit health-care facilities; however, they must be in rural, underserved areas.

These health centers provide care to a diverse population and have even shown reduced mortality in treatments compared with hospitals (Wennburg et al., 1998). For FQHCs and FQHC Look-Alikes, in particular, over two-thirds (68 percent) of patients who seek care have patient incomes at or below the poverty level, 22 percent are uninsured, and 47 percent are covered by Medicaid. More

TABLE 4-1 How Specific Community-Engagement Opportunities Can Benefit Research Organizations and Communities

Mission	Community-Engagement Opportunity	Benefit to Community	Benefit to Academic Research Organizations
Research	Scientists, regardless of discipline, develop research questions in collaboration with community.[a]	Aligns research resources with local needs; increases connection to STEM mentors and training; develops community capacity to use research, seek grants, and increase community-based organization's sustainability; and ensures data can be used to support local advocacy efforts.	Increases relevance of research and likelihood that findings will be broadly implemented; increases recruitment and retention in clinical studies; enhances scientists' competitiveness by strengthening external validity; increases internal validity by adding community perspective to construct definitions and measurement tools or strategies; produces stories useful for marketing and advocacy; and develops trainees' skills in communication, collaboration, and engagement.
	Researchers work with community members to improve the relevance and conduct of studies, as well as the dissemination of findings and discoveries.		
	Research centers invite community members to serve on search committees and interview faculty applicants, and incorporate those perspectives into hiring decisions.	All of the above, plus provides the community the opportunity to exercise agency and influence decisions and increases opportunities for mutually beneficial projects.	
Education	Educators integrate the community and community health needs assessments when developing interprofessional learning opportunities. Community-based learning is evaluated in terms of outputs and outcomes relevant for learners, community members, and the research organization itself.[b]	Ensures learner service aligns with community needs in respectful and valued ways; evaluation allows improvement to community-based organization's program and exposure to evaluation science, which is important for the partner agency's own improvement efforts; and learners passion and commitment present a different side of the health-care system.	Develops interprofessional competencies; develops trainees' communication, collaboration, and engagement skills; exposes learners directly to local sociocultural contributors to health; and produces stories useful for marketing and advocacy purposes.

Mission	Community-Engagement Opportunity	Benefit to Community	Benefit to Academic Research Organizations
	Learners across health professions directly contribute to local community health needs assessments processes as data collectors or analysts, or by presenting results to community groups.	Increases exposure and connection to learners, increases awareness of local health-improvement activities, and presents more opportunities to codesign community health needs assessments-related health interventions.	Provides additional labor for teaching hospitals' community-related administrative functions; provides research practicums focused on survey design, focus group development and execution, data analysis, data reporting, program development, etc.; offers educators new opportunities to teach about social determinants of health, population heath, public health, etc.; and provides graduate medical education involvement and contributes to instruction on health and health-care disparities.
	Program directors routinely model the stratification of their patient and participant data by sociodemographic characteristics to identify health-care inequities. Trainees partner with community members, patients, and faculty to develop interventions.	Results in improvements to work flows more likely to benefit patients' and community members' health outcomes.	Contributes to instruction on health and health-care disparities; targeted disparity-focused quality improvement efforts can have an effect on overall measured quality; when implemented in an accountable care organization or similar setting, can result in increased shared savings; advances scholarly output; and increases trainees' patient and community-engagement skills.

continued

TABLE 4-1 Continued

Mission	Community-Engagement Opportunity	Benefit to Community	Benefit to Academic Research Organizations
Clinical care	Clinical teams use data across multiple levels—clinical, sociodemographic, and neighborhood—to tailor care plans in ways that are responsive to the health and the environmental or social profiles of their patients.	Improves health outcomes, enhances knowledge of and access to community assets, and increases demand or support for local community-based organizations' programs.	Improves quality of care, particularly on measures related to readmissions, cost, and resource use; enhances physician and provider wellness through increased ability to manage patients' social factors; increases efficiency and effectiveness of hospital community health or prevention efforts by enhancing alignment or reducing redundancy with local initiatives; and advances scholarly output.
	Clinicians and care teams, through their electronic health records, have robust linkages to hospitals' community health-improvement efforts and make appropriate and timely referrals to community assets that can provide social support and resources for patients and their families.		
	Care team members spend time at community-based referral partners meeting staff, engaging patients, and learning about local social service processes to improve their community knowledge and profile and to increase their ability to make appropriate, knowledgeable referrals.		

[a] Joosten et al., 2015; Kost et al., 2017.
[b] Guthrie et al., 2016.
SOURCE: Table adapted from Wilkins and Alberti, 2019.

than 28 million patients received care at 1 of 1,375 FQHC delivery sites in 2020 (an additional 679,000 were served at 87 FQHC Look-Alikes in 2020), with FQHCs and FQHC Look-Alikes treating nearly one in seven uninsured people in the United States (HRSA, 2020a, 2020b). Greater than 62 percent of FQHC and FQHC Look-Alike patients are members of racial/ethnic minority populations, including 37 percent who are Latino and 26 percent who are African American. More than one-quarter (28 percent) of patients are best served in a language other than English (HRSA, 2022). Similarly, rural health centers serve more than 8 million people across 4,400 delivery sites in 45 states (NARHC, 2022). As such, community health centers are an ideal setting to recruit diverse participants into clinical trials and research. This research activity, however, will require infrastructure and support for the community health centers.

Nevertheless, the barriers to clinical trials and research recruitment at community health centers are multifactorial. Health-care providers who work in community settings outside of academic centers may have limited knowledge about available research opportunities. This may be particularly true in rural communities (Paskett et al., 2002). The same is even true for physicians near academic medical centers: in a survey of more than 100 physicians in New Jersey, lack of awareness of cancer research opportunities was reported by 95 percent of PCPs, 84 percent of non-oncology specialists, and even 50 percent of oncologists (Hudson et al., 2005).

While electronic health record (EHR)–based prescreening of patients for trial eligibility has been a common practice in many health systems (Canavan et al., 2006; Sullivan, 2004; Wilcox et al., 2009), it poses challenges for many health systems, especially those lacking sufficient EHR infrastructure. Some clinics may be unable to successfully query the EHR using study inclusion and exclusion criteria. Often EHRs have heterogeneous data structures that can make it difficult to consistently apply study inclusion and exclusion criteria across sites (for multisite studies) (Hersh et al., 2013; O'Brien et al., 2021). Health centers that lack onsite specialty care services, may inadequately track the delivery of clinical services completed offsite. Many of these problems are more pronounced at community health centers, which may have limited data infrastructure and fewer staff trained to carry out research functions. Initiatives such as the Community Health Applied Research Network are working to address some of these infrastructure challenges and may serve as a model to expand and grow capacity for FQHCs (see Box 4-1).

Individuals who receive care at community health centers may frequently change their address or phone number or may be houseless, creating obstacles to being recruited for study participation. These individuals also may face competing demands, such as work or caregiving responsibilities, or may lack transportation to attend research-related appointments. Patients' health insurance coverage, which may be inconsistent or variable, may present barriers to obtaining the needed clinical services required for study participation. A high participant no-

> **BOX 4-1**
> **Community Health Applied Research Network**
>
> The Community Health Applied Research Network (CHARN) was established in 2010 to serve an estimated 1 million people from a range of underserved populations such as those of low socioeconomic status, those who do not have access to health insurance, or racial/ethnic minority groups. This network, funded by the Health Resources and Services Administration, consists of 18 federally funded community health centers, with four research nodes and one national data coordinating center.[a, b] The goal of CHARN is to improve patient care at this network of federally funded health centers by developing and refining clinical data systems; creating infrastructure to better collect patient data across health centers; training health center personnel in research methods and protocols; improving translation of research findings into effective, patient-centered clinical practice; and fostering collaboration among care teams and other CHARN health centers.[b] CHARN centers also work to develop proposals to obtain additional funding through the federal government to execute these objectives. CHARN was expanded in 2014 to include several new initiatives. For example, researchers interested in patient-centered outcomes can now access patient data that had been otherwise unobtainable for "out-of-network" researchers. Further, the U.S. Department of Health and Human Services' Office of the Assistant Secretary for Program Evaluation (ASPE) has also worked to increase data infrastructure by making all patient data from clinical visits available from 2006 to 2013, an increase from 2008 to 2010.[c] ASPE has also released several publications on the success of CHARN related to increasing health data infrastructure and patient-centered health outcomes.
>
> ---
>
> [a] See https://www.ncbi.nlm.nih.gov/pmc/articles/PMC4371501/.
> [b] See https://www.kpchr.org/CHARN/public/index.aspx?pageid=1.
> [c] See https://aspe.hhs.gov/strengthening-expanding-community-health-applied-research-network-charn-registry-conduct-patient.

show rate and the need to translate study materials into multiple languages (and/or enlist interpreter services) may impose additional study costs.

Specialty associations, such as the Association of Black Cardiologists and the Association of Black Gastroenterologists and Hepatologists, among others, can serve as effective organizations to promote clinical trial recruitment in traditionally underrepresented population groups (Ofili et al., 2019).

Drug and Device Companies and Clinical Trial Recruitment Centers

As health care evolves toward precision medicine, it is essential that the biologic differences among populations—and how these differences affect pathology, response, tolerability, and outcome—are comprehensively investigated in the context of clinical trials. Pharmaceutical and device companies have an essential role in developing and implementing successful strategies, measurable outcomes, and robust outreach plans to include diverse populations efficiently and effectively in clinical trials.

Overly restrictive study design, stringent eligibility criteria, and continuous activation of clinical trials in sites based on their academic prominence or speed of enrollment often has resulted in the exclusion of underserved patient populations (much to the detriment of inclusive research). This has contributed to the widening disparities between patients who are expected to benefit from the new research in day-to-day clinical practice. It is clear that eliminating the factors and problems that limit trial participation would improve the generalizability of results. Problems that prevent the inclusion of diverse populations in industry-funded clinical trials include patient out-of-pocket costs, which are often not covered in the informed consent process; industry pressures to gather data quickly; and the selection of easy-to-recruit samples being incentivized (Iltis, 2004). Payment structures often pay per participant, which further incentivizes institutions to focus recruitment on populations that are easiest to recruit. Although many of these problems are not unique to industry-sponsored trials and are present in federally funded research as well, most clinical trials are industry-funded and the business demands of industry make these problems particularly acute.

Broader Landscape

Institutional Review Boards[2]

All research that involves human subjects must be reviewed and approved by an institutional review board (IRB) (see the Common Rule, 45 CFR 46).[3] IRBs are charged with protecting the rights and welfare of human subjects who participate in research. The evaluation of human subjects' rights and welfare is guided, in part, by key ethical principles established in international and national guidelines such as the *Declaration of Helsinki*) and the *Belmont Report: Ethical Principles and Guidelines for the Protection of Human Subjects of Research* (National Commission for the Protection of Human Subjects of Biomedical and Behavioral Research, 1979; WMA, 2008). The *Belmont Report*, issued in 1979, was commissioned by law in response to the abuse of human subjects in the U.S. Public Health Service Syphilis Study at Tuskegee (Brandt, 1978). The report explores the boundaries of medical research, the determination of risk versus benefit in research, the appropriate selection of human subjects for participation in research, and the fundamentals of informed consent. Importantly, it also

---

[2] This section relates to IRBs that exist under Food and Drug Administration regulations. It is important to acknowledge that research done on tribal lands falls under the individual tribes' IRBs, because tribes are sovereign nations. See Kuhn et al., 2020.

[3] See also Consideration of the Principle of Justice under 45 CFR part 46, July 22, 2021, at https://www.hhs.gov/ohrp/sachrp-committee/recommendations/attachment-a-consideration-of-the-principle-of-justice-45-cfr-46.html.

outlines key ethical principles to guide research with human subjects. These ethical principles include (1) respect for persons, which refers to the right to self-determination, or autonomous decision-making; (2) beneficence, which refers to the obligation to protect the well-being of human subjects; and (3) justice, which refers to the fair distribution of the benefits and burdens of research participation.

The ethical principles outlined in the *Belmont Report* are operationalized through the day-to-day work of IRBs, which operate under the guidelines and administration of the Health and Human Services (HHS) Office for Human Research Protections (OHRP). The Code of Federal Regulations (45 CFR 46) guides the structure and function of IRBs, and particularly, the IRB review process. According to the Code of Federal Regulations, IRBs must have at least five members, and those members must have sufficient knowledge or experience to evaluate research activities proposed by investigators affiliated with the institution. IRBs are charged with ensuring that risks to human subjects are minimized through the use of sound scientific processes and their review focuses on the following key elements of research proposals: risks and benefits to human subjects, safety, protections of privacy, equitable selection of human subjects, and informed consent, with particular attention to coercion and undue influence. The latter element of IRB review can present barriers to enrolling excluded and underrepresented populations.

The ethical principle of respect for persons is operationalized in the research consent process, which is meant to support participants' right to autonomous decision-making, and protect participants with diminished capacity for self-determination. The *Belmont Report* describes capacity for self-determination as fluid—increasing with maturity, but potentially lost in some natural and social circumstances such as severe illness, cognitive disability, or restricted liberty. The Code of Federal Regulations adds more clarity to this idea by its identification of groups whose vulnerability demands increased protection beyond those afforded to all human subjects in research. Those groups include children, prisoners, persons with impaired decision-making capacity, and economically or educationally disadvantaged persons (45 CFR 46). The code describes these groups as vulnerable to coercion and undue influence in research participation, and therefore directs IRBs to pay particular attention to their consent to research. The code does not, however, define coercion or undue influence, thereby leaving the interpretation to IRBs, who have largely focused on the potential for compensation and incentives to be unduly influential or coercive (Largent and Lynch, 2017).

Most research with human subjects involves some form of compensation for participation. Forms and amounts of compensation—sometimes referred to as incentives—vary by study, with one-time surveys and interviews typically offering smaller incentives compared with lengthy clinical trials that involve medical interventions and frequent study visits. IRBs tend to lean toward viewing higher payments as coercive, and err on the side of keeping payments low (Largent and Lynch, 2017). However, coercion requires "the overt threat of harm" to gain another person's compliance (DHEW, 1979). Thus, some ethicists argue that research payments cannot be considered coercive, which is a perspective articulated by the director of HHS OHRP (Largent and Lynch, 2017; Meeker-O'Connell and Menikoff, 2021). Undue influence refers to an offer that encourages the potential recipient to do something that is unreasonably against their best interests or values (Emanuel, 2005). It does not refer to an offer that encourages the potential recipient to do something reasonable that they might not do in its absence. IRBs cannot approve studies that pose unreasonable risk to potential participants, which means that any approved study should be considered a reasonable undertaking for its target population, on the whole. Thus, it is difficult to argue incentives are a form of undue influence; yet, research indicates IRB members are concerned about coercion and undue influence when substantial payments are offered to research participants (Largent et al., 2012).

IRB members' concern about coercion and undue influence in the form of incentives reflect their commitment to the canonical principle of respect for persons. However, limiting incentives may ultimately compromise other equally important principles, including beneficence and justice. Some research provides direct benefit to participants, thereby supporting their well-being. In the absence of sufficient payment or other supports (e.g., food, transportation, childcare), persons who might benefit from research participation are prohibited from doing so. The people most likely to bear an excess economic burden of research participation, especially in the absence of substantial support, are those who are in hourly jobs, or live far from academic research centers, or have dependents for whom they must provide care (Nipp et al., 2016). Excluded and underrepresented populations are more likely to be in these social circumstances. Thus, without adequate support, their ability to participate is restricted, they miss opportunities to enhance their well-being, and the distribution of research benefits and burdens is unjust. The underrepresentation of particular demographic groups also limits the opportunity to generate sufficient data on the safety and efficacy of new therapeutics for them; this may create injustice in delayed access to interventions, or in unforeseen differential outcomes (Hume et al., 2017; Knopf et al., 2020). Although not the focus of this report, it is also important to recognize the negative impact IRBs can have on the enrollment of adolescents in clinical trials and clinical research. Since many IRBs require guardian consent, this may disallow adolescent participation if a parent or guardian is not comfortable with or is distrustful of clinical trials and clinical research. Allowing adolescents to make independent decisions on whether they would like to enroll in a clinical trial or not may reduce barriers to enrollment and further understanding of health disparities (e.g., sexual health or substance use) in adolescent populations (Fisher and Mustanski, 2014; Fisher et al., 2021; Gilbert et al., 2015; Knopf et al., 2017).

Although IRBs certainly have a role to play in increasing the representation of excluded and underrepresented populations in clinical trials and clinical research, these bodies are focused on protecting individuals, not communities. Engaging with community advisory boards offers an opportunity for researchers to anticipate and address community concerns and to help communities understand the risk of the proposed research (Quinn, 2004; Strauss et al., 2001). CABs can also facilitate the involvement of community members on local IRBs, offering additional protections to community members and helping to alleviate issues of trust.

Funders

Research funders have several roles and responsibilities that can influence the diversity of clinical trials. Traditionally, funders' roles include prioritizing research topics, approaches, and methods; receiving and evaluating grant applications; selecting suitable proposals for funding; and evaluating the output of the research (Brantnell et al., 2015; Kessler Foundation, 2011). In each of these stages, funders have opportunities to promote diversity, but they also face constraints that may limit the effectiveness of efforts to enhance representativeness.

Funding Priorities. Funders set and implement research funding agendas that can ultimately affect the clinical research that is conducted and the scope of these projects. Sharing this agenda through published criteria and grantee informational sessions provides opportunities for emphasizing the ethical, scientific, and clinical importance of diversity in clinical trials. Moreover, the research agenda is typically informed by few scientists from underrepresented groups. Within the National Cancer Institute's (NCI) Intramural Research Program, for example, only 1 percent of senior investigators (those granted tenure by the deputy director for intramural research) are Black and 2 percent identify as Hispanic. There are no Black and Hispanic senior scientists and clinicians (managers of large institutes' or centers' research departments) at the NCI. Instead, three-quarters or more of the NCI's senior scientists and investigators are white. Nearly two-thirds of R01s are awarded to white applicants, with Black scientists and Hispanic scientists making up only 1 percent and 5 percent, respectively, of awardees (Ong, 2021). Further, although the NIH provides diversity supplements to investigators to support a diverse and inclusive workforce, many are limited to 2 years for the training of junior investigators, which means very little time for establishing partnerships, recruitment, and retention for projects.

Funders can also prespecify diversity targets for the research studies. This approach has been successful in several large research studies, including the NIH Diabetes Prevention Program and Systolic Blood Pressure Intervention Trial, or SPRINT (Group, 2015; Knowler et al., 2002). SPRINT, for example, which examined blood pressure in 9,361 people, set specific recruitment targets and ensured that trial sites were diverse and could bring in diverse patients to achieve recruitment goals (Ambrosius et al., 2014; Greer, 2015).

Proposal Reviews. Through the review process, funding agencies can give priority to projects that include sufficient numbers of underrepresented persons. The NIH has implemented initiatives designed to foster the inclusion of underrepresented groups in NIH-supported clinical research trials and to incorporate valid analyses by sex and gender (NIH, 2001b, 2017a). Federal funders, such as the NIH, National Science Foundation, and Patient-Centered Outcomes Research Institute, require enrollment tables that are incorporated as part of the review process. The weight that reviewers give data may vary, however, and can be applied inconsistently. Further, these enrollment tables are not part of the score-driving criteria, which limits the impact they have on funding decisions.

In addition to review criteria, those reviewing grant applications and making funding decisions influence the type of research that is carried out. Studies have shown that NIH study sections, which review and decide which clinical research grants get funded, are overwhelmingly white. According to one study, 2.4 percent of study section members in the period FY 2011–2015 were African American/Black compared with 77.8 percent who were white (Hoppe et al., 2019). The Center for Scientific Review at the NIH is tasked with improving disparities in peer review and has stated that "there must be diversity with respect to the geographic distribution, gender, race, and ethnicity of the membership of study sections" (NIH, 2020b). It is important to acknowledge the availability of these data that illustrate the lack of diversity among academic medical centers and study sections for publicly funded biomedical research. Moreover, data on investigators in industry and other private entities are not publicly available.

Funding for Recruitment and Retention. Recruitment and retention of diverse participants can be costly; one study estimates recruitment costs ranging from $129.15 to $336.48 per enrolled patient (Penberthy et al., 2012). Recruitment can require higher staffing levels, more frequent contacts, longer accrual periods, additional funding, and more flexible funding to enhance trial accessibility for low-income participants, those with caregiving responsibilities, workers without flexible hours, and individuals with other competing priorities. For example, transportation to trial sites is often a deterrent to participation of underrepresented populations, and need for childcare can limit participation of caregivers, who are more likely to be female. Collaboration with community organizations to colocate services in community venues, such as faith institutions, or provide mobile services may increase the reach and effectiveness of the clinical trial recruitment and retention efforts. Many funding agencies often underestimate the increased effort and financial resources needed to ensure diversity in research studies. Flexible funding that can be used to promote or augment these strategies can play a critical role in increasing trial diversity and warrants consideration by IRBs (see Institutional Review Boards section, above).

Post-award Reporting and Monitoring. Review and examination of participant accrual and review of adverse events is a routine role of funders as well as the Food and Drug Administration and individual data safety monitoring boards. In this phase, systematic, timely, and transparent collection and reporting of trial diversity metrics is a requirement for intervening to modify trial protocols to promote more inclusive recruitment or avoid differential disenrollment (Artiga et al., 2021).

Evaluating the Output of the Research. In addition to clinical effectiveness outcomes, funders often examine the impact of their research using metrics such as publications and patents produced. Making the diversity of participants that are recruited and retained an explicit outcome to be evaluated and reported can be an important strategy for enhancing patient and community trust in the research process, increasing the applicability of the research findings to women, minority communities, and older adults, and influence payers' and providers' acceptance of the findings for groups who were not adequately represented in the research.

Medical Journals

In many ways, medical journals serve as the gatekeepers to medical knowledge, holding the key to publishing studies that advance clinical practice and improve health. Thus, journals yield great power along with accountability for what is and is not published in their pages. In the past year, leading medical journals have acknowledged that "they must do better towards inclusion and antiracism in all journal related activities," and many have issued initiatives and calls to action to increase diversity, equity, and inclusion. In October 2021, the *New England Journal of Medicine* announced it would begin requiring authors to submit a supplementary table describing the disease or health problem under study, its distribution in the population (e.g., by race, ethnicity, and sex), and representativeness of enrolled study participants (NEJM, 2021). This is an important step, but whether it becomes a significant factor in determining the acceptance of manuscript submissions remains unclear. Further, journals still have a long way to go, as reflected by representativeness of editors and by their rate of publications in these areas.

Regarding representation, among the 346 editors and editorial board members across *JAMA* and the JAMA Network journals, 71 percent are white, 19 percent Asian, 6 percent Black, and 4 percent Hispanic; 38 percent are women (Fontanarosa et al., 2021). In a review of 444 leading medical journals, women represented only 21 percent (94) of editors in chief (Pinho-Gomes et al., 2021), and this rate has changed little over the past decade (Jacobs et al., 2021). Of 215 leading surgery journals, only 7 percent of editors are women (Kibbe and Freischlag, 2020). For publication rates on issues of diversity or health disparities, a recent review indicated that the proportion of articles on these topics relative to all articles published was only 7 percent at the *Journal of General Internal Medicine* and was less than 2 percent at other leading general medicine journals (Jackson et al., 2021). Moreover, few papers in any journal addressed "racism" in their title, abstract, or key words (Jackson et al., 2021; Rhea et al., 2020).

While some journal editors have expressed skepticism about the power of journals, and post hoc publication, to influence inclusion in clinical research, there is in fact evidence to suggest that journals can exert a large impact. The requirement of adherence to reporting standards, such as CONSORT (Consolidated Standards of Reporting Trials) (Moher et al., 2010) and other standards, and mandatory trial registration, such as ClinicalTrials.gov, have standardized and raised the quality of research design, reporting, monitoring, and transparency for clinical trials. There is no question that similar influence could be exerted with the requirement for mandatory standards and reporting on diverse inclusion in clinical studies. Moreover, journals can influence diversity and inclusion across multiple domains (Rivara et al., 2021). Ultimately, a unified and concerted effort by medical journals, such as through the International Collaboration on Standards and Policies through the Royal Society of Chemistry's Joint Commitment for Action on Inclusion and Diversity in Publishing (RSC, n.d.), may pave the way for ongoing and long overdue change.

5

Facilitators of Successful Inclusion in Clinical Research

The analysis draws substantially from the research paper by Franchesca Arias, Ph.D.; Nicole Rogus-Pulia, Ph.D., C.C.C.-S.L.P.; and Amy J. Kind, M.D., Ph.D., which was commissioned for this study. See Appendix C for the full research paper.

There is substantial quantitative data demonstrating the size and scope of the problem of underrepresented and excluded populations in research; however, there is a dearth of critical qualitative data about facilitators of successful representation in clinical research. The experiences of research teams who have successfully enrolled diverse participants contribute to a better understanding of the facilitators that can be leveraged to make progress. These data are needed to develop a robust science of inclusion that can help the field evaluate and scale effective, real-world engagement and recruitment strategies. To that end, this chapter provides an overview of evidence on sentiments, facilitators, beliefs, and attitudes from study investigators, staff, and participants for overcoming barriers to the inclusion of women and underrepresented minorities outlined in Chapter 4.[1] It highlights key themes and facilitators that have demonstrated effectiveness to enhance recruitment and retention of diverse populations in clinical studies. In each section, the findings include reports from 20 qualitative interviews conducted in 2021 with research teams (investigators and staff) involved in clinical trials who successfully achieved diverse enrollment. The research teams were identified using a systematic process to ensure that therapeutic areas were equally represented. (The next section summarizes the study approach; Appendix C con-

---

[1] The committee defines facilitators as strategies and factors that facilitate success in overcoming barriers to the increased representation of women and racial and ethnic minority population groups in clinical research.

tains the full analysis). These qualitative data are bolstered by evidence from the literature reviewed by the committee. Finally, this chapter summarizes practical and innovative facilitators, particularly those that may be replicable and scalable in future studies.

INSIGHTS INTO EFFECTIVE FACILITATORS AND STRATEGIES FOR INCLUSION

The qualitative evidence for this chapter is largely derived from a mixed-methods study that the committee requested be commissioned for this report (see Appendix C). The purpose of the study was to characterize current efforts on representativeness in clinical research and to systematically identify and describe recruitment and retention strategies that can contribute to more diverse clinical trial participant populations. Individual comprehensive interviews were conducted with research team members (an investigator or coordinator) with experience recruiting underrepresented groups. Twenty interviews were completed in 2021 that focused on understanding facilitators to recruitment and retention into clinical trials.

At the beginning of the study, the authors first identified the six diseases associated with highest mortality in the United States (heart disease, cancer, chronic lower respiratory disease, stroke, Alzheimer's disease, and diabetes). Next, a systematic review was conducted to identify published clinical trials in these six therapeutic areas between 2001 and 2021 that successfully recruited diverse populations, defined as having at least 50 percent or higher enrollment of the county-level base rate, the state-level base rate, or the national-level base rate (for single-site studies, county-level data were used; for multisite studies, state-level data or national-level data were used, depending on whether sites were within the same state or dispersed across the United States) in at least one of the three categories of sex, race, and ethnicity mandated by the National Institutes of Health (NIH). From more than 130,000 trials that were identified, 162 trials stratified by disease and geographic location were selected. Of these, 142 trials met criteria for diverse enrollment (i.e., at least 50 percent or higher enrollment of at least one of the three NIH-mandated categories of sex, race, and ethnicity), and were invited to participate in interviews. Notably, less than 33 percent of these trials reported information about ethnicity, and less than 66 percent of trials included categories of racial/ethnic representation. Research team members (an investigator or coordinator) from each of the 142 identified studies were invited to participate in a qualitative interview (see Appendix C for full study details).

Based on 20 completed in-depth qualitative interviews with rigorous thematic analysis, 8 major themes emerged, which provided insights into key facilitators to inclusion. These themes are (1) starting with intention and agency to achieve representativeness; (2) establishing a foundation of trust with study participants and community; (3) anticipating and removing barriers to study participation; (4)

adopting a flexible approach to recruitment and data collection; (5) building a robust network by identifying all relevant stakeholders; (6) navigating scientific, professional peer, and social expectations; (7) optimizing the study team to ensure alignment with research goals; and (8) attaining resources and support to achieve representativeness. Table 5-1, at the end of the chapter, provides an overview of strategies to enhance inclusion derived from the interviews, organized by theme. Further details on each of these themes are described below.

Starting with Intention and Agency to Achieve Representativeness

From goal setting to community partnering strategies, intentionality and planning are critical themes for overcoming the systemic barriers previously outlined to the inclusion of underrepresented minorities and women in research (McMurdo et al., 2011). While planning and engagement with diverse communities is resource, time, and labor intensive, it is critical to advancing inclusion. According to research teams that participated in the analysis, "It's a lot of work and a lot of time and it takes years. . . . We've been working with the same community partners now for 12, 13 years. They see us all the time." They emphasized that a multistage process is required to achieve representativeness and that contact with communities should begin long before recruitment starts and extend long after the study ends.

Evidence suggests that to build relationships with the community, research should continue to affect changes in communities long after the study ends and throughout all stages of the study process (Gluck et al., 2018). Research teams emphasized that collaboration with community members specific to recruitment and retention strategies occurs across different stages of the study. For example,

> I think some of the principles that are laid out for stakeholder engagement are basically to involve them in the design of the study, the conception of the study, what questions you're asking, as well as in how you're doing, the recruitment, who you're recruiting, what your materials are, and then what the study involves, like kind of soup to nuts kind of thing. And so I try to do that as much as I can.

Research teams reported that being intentional about having representation of historically underrepresented groups was instrumental to their success. Setting a priori recruitment goals for the inclusion of underrepresented groups is essential to planning and can help research teams measure progress and develop more effective engagement strategies (Javid et al., 2012).

Research teams also emphasized the importance of considering access barriers and the lived-realities of study participants in research design. For example,

> It's absolutely important in terms of behavioral interventions and how you implement [with] certain people or not if you don't have access to the things that people of high socioeconomic status take for granted. If you don't have that kind of access, then you're not going to be as able to implement any intervention, especially behavioral ones that require changes in lifestyle, taking time out of your day and stuff like that.

Research suggests that prioritizing access to health-care resources can facilitate research inclusion for underrepresented communities (George et al., 2014).

Research teams also reported that intentionality is sometimes driven by external factors, such as requirements by funding agencies, the need to recruit from a given state or setting, and characteristics of the diseases, such as rates in diverse populations.

Establishing a Foundation of Trust with Participants and the Community at Large

Building and maintaining trust with both study participants and their larger communities is foundational to achieving equity in research (Barnes and Bennett, 2014). Research teams reported that the history of abuse in prior studies, experiences with other research groups that approached underrepresented communities for the purposes of a study and who did not remain engaged, and beliefs that research is not beneficial to the community are critical barriers to establishing trust with persons from diverse communities.

The development of trust requires a long-term commitment by principal investigators, study teams, and local institutions involved in the research. Building trust over time takes consistent engagement in the community beyond the confines of the study itself, developing meaningful relationships with study participants, and giving to the community without the expectation of anything in return (Kretzmann and McKnight, 1993). Research teams emphasized that while trust has to be built over time, trust can be broken with individuals and communities in an instant. For example, "There's such trust building, that . . . takes a while. And if one person drops and doesn't keep the trust, then I'm not going to be able to most likely get back that location again."

Developing robust community partnerships that are equitable and not hierarchical in nature can mitigate distrust in communities and can help research teams effectively leverage resources for truly meaningful and translatable work in partnership with community members (Waheed et al., 2015). According to one research team member,

> I think that is the goal to get to full equity with the community partner, writing the grants and getting the money and sharing everything from the ground up to the study. I think we're still unequal with academic partners. So doing a grant writing, getting the funding and working with community partners and giving them funding from the grant. So I think there's still this hierarchy, unfortunately. We're trying to break those down. We're trying to get to parity as much as possible. And that's just going to take time and it's going to take investment.

In addition to facilitating recruitment, establishing relationships with community leaders provided opportunities to understand the needs of the community in order to build trust over time.

Anticipating and Removing Barriers to Study Participation

To assure accessibility to study participation for members of underrepresented communities, anticipating and removing barriers to participation—which are described in Chapter 4—is critical. For example, one systematic review of 44 articles found that facilitators to research participation included tailoring recruitment strategies to each community group (George et al., 2014). Recognizing heterogeneity within cultural groups is key, and a one-size-fits-all approach will not work. Investigators should take an individualized approach, without compromising the science, to create protocols that allow for and acknowledge individual experiences. For example, according to one research team member interviewed, "There was no cultural tailoring at all. There was a ton of individualized tailoring. The intervention itself is highly individually tailored. And so we just developed personalized approaches to everyone. And, in doing that, we didn't have to put people into categories to try to tailor to them." Other important solutions include collaboration with interpreters to provide services to non-English-speaking prospective study participants and/or providing options for in-home or remote visits to overcome linguistic or physical access barriers. For example implementation of the asset-based community development, or ABCD, approach, which includes a high level of community assessment, engagement, and involvement before actual recruitment begins, in the Wisconsin Registry for Alzheimer's Prevention (WRAP) led to a 400 percent increase (from 0.8 to about 8.0 percent of the study sample (131 of 1,573) in the participation of African Americans (Green-Harris et al., 2019). These themes are also discussed in more detail in Chapter 2.

In Clark et al. (2019), study participants noted several strategies that clinical research staff could employ to overcome barriers and improve participation and retention in clinical trials, including "…rapport with participants; attentiveness and sensitivity to patients' concerns or needs; flexible scheduling to accommodate participants after hours and on weekends; post visit follow-up telephone calls to assess participants' well-being and address any concerns; and regular touch-base contacts with participants."

Adopting a Flexible Approach to Recruitment and Data Collection

Research teams from the successful studies recognized the importance of flexibility to enhance recruitment and retention of diverse groups. Research teams frequently described recruitment strategies adapting and evolving as studies progressed. Recruitment techniques were incorporated or abandoned in response to study needs, and changes were guided by input from community representatives and other relevant stakeholders. This adaptability extended beyond recruitment. For example, flexibility at the time of data collection was reported as necessary to retain study participants, particularly those with limited resources or constraints on their time due to competing demands such as childcare or eldercare. For example,

So we had to be very flexible in how we collect the data. We ultimately ended up giving people multiple data collection options, so we tried to enroll everyone and do baseline data collection in person for folks, for literacy reasons, for understanding comprehension and for trust building. And then after that, they could meet us in person or in the clinic. They could meet us in person in our research offices. They could do it online via REDCap. They could do it via phone with a research assistant. They could be mailed a paper survey. And similarly, they could go in for a . . . test at a clinic or they could do a mailing kit.

Flexible approaches meeting study requirements were instrumental in the success of diverse enrollment. Several prior reports demonstrate this phenomenon. For example, one qualitative analysis of interviews with 30 Native Hawaiian women identified that disseminating study information through community channels with targeted outreach to religious and social organizations as well as face-to-face contact with researchers in a culturally tailored way would help with recruitment and retainment (Ka'opua et al., 2004). Another example comes from a recent randomized trial of a mobile health support program for diabetes self-care that utilized multiple retention strategies for minority populations. The strategies included flexibility in participation (e.g., multiple methods for data collection), communication (e.g., tracking contacts), and community building (e.g., study branding and newsletters). With these flexible and multipronged approaches, retention was greater than 90 percent at each follow-up assessment that occurred over 15 months (Nelson et al., 2021).

Recent U.S. Food and Drug Administration (FDA) guidance (Clark et al., 2019; FDA, 2020c) recognizes the need to make trial participation less burdensome to enhance recruitment, and supports the use of flexible approaches to reduce the frequency of study visits, to build in flexibility in visit windows, to consider electronic communication or digital health technology tools to replace site visits, and to consider the use of mobile staff to conduct study visits in study participants' homes.

Building a Robust Network by Identifying All Relevant Stakeholders

Research suggests that engaging in mapping to identify all the relevant stakeholders in a community can help study teams develop more equitable study designs and identify individuals and organizations that can help drive the recruitment and retention of diverse study participants (Larkey et al., 2009).

According to the research teams that were interviewed, identifying these stakeholders and their level of needed involvement varied based on cultural preferences of the prospective study participants, the condition being studied, and the nature of the research study. The term *stakeholder* was defined broadly to include caregivers, family members, friends, clinical providers and administrators, community advocates, peers, religious leaders, and political figures.

Strategies for consistently engaging communities such as community advisory boards can help inform protocol development and study execution (Buck et al., 2004), whereas specific stakeholders, such as community health workers and patient navigators, have been found to help drive the recruitment and retention of underrepresented groups in research (Choi et al., 2016). Studies consistently show that community health workers, who typically focus on informing patients about the importance of adherence to a particular healthy behavior and who patients can go to for help, support, and informal counseling, improve outcomes for patients. For example, in one study of community health workers in the Bronx (a borough of New York City), adding community health workers to a medical home led to a decline in emergency department visits and hospitalizations among patients with chronic health conditions (Findley et al., 2014). In another study, community health workers helped improve recruitment and retention of immigrant women in a randomized trial to promote mammograms and Pap tests (Choi et al., 2016). Patient navigators, who typically handle patient problems as they arise, may be inserted into health-care studies to help patients adhere to recommended care (Dohan and Schrag, 2005). In one of the foundational studies examining the effectiveness of patient navigators in expanding cancer screening and care in medically underserved populations, Freeman et al. (1995) found that patients who had a navigator were far more likely to complete recommended breast biopsies and do so in far less time than those without navigators. In fields where use of patient navigators is more common, such as cancer screenings and care, patient navigators help to catch disease at earlier stages, help ensure patients show up to follow-up appointments, and help ensure patients receive follow-up care once they have a diagnosis (TCFHA, 2012).

Important themes emerged related to patient and caregiver engagement. For example, developing relationships with caregivers and family members was identified as instrumental to recruitment and retention of underrepresented groups. According to one participant, "I realized that not talking to caregivers was a pretty big misstep in our original trial. If you have these populations that are vulnerable enough to have caregivers and other people who are already kind of with them maybe consider including them as part of the trial and obviously with patient consent, sort of incorporating it." Further, conceptualizing study participants as partners in research was highlighted as important and requiring openness and flexibility by the study team to learn from the study participants' experiences.

Navigating Scientific, Professional Peer, and Societal Expectations

Research teams described challenges related to scientific and societal expectations, which sometimes conflicted with maintaining scientific rigor. Many of the research teams perceived that efforts to promote representativeness, and decisions made to support these efforts, are not fully embraced or supported by colleagues and organizations responsible for making funding and/or budget decisions. Creative strategies designed to engage communities that have traditionally been underrepresented in research are often not valued relative to more traditional strategies, which tended to involve rigid protocols applied within standard working hours (e.g., 9 a.m.–5 p.m.), conducted onsite, and carried out by staff who were not multilingual. These traditional approaches to retention and recruitment may be burdensome for prospective study participants with multiple vulnerabilities, and may result in less participant diversity. Thus, providing a more flexible infrastructure (e.g., more flexible protocols, off-hours participation, offsite participation including by remote or in-home means) may be critical to enhancing participant diversity.

Research teams expressed concern that the current emphasis on recruitment and retention of diverse study participants contrasts with consistent underfunding of disparities researchers despite the additional costs needed to conduct diverse enrollment in all research studies. For example,

> It seems that there's a real incongruence where the NIH is saying disparities work, disparities work, disparities work, and then you put it in and reviewers don't acknowledge the disparities aspect. They are fixated on errors in your approach or concerns about your theoretical model, and so it does seem that there is an incongruence in the way that the funding source of NIH wants to value efforts to recruit and retain these folks and then the way that it's reviewed. So that is an issue.

It is well documented that scientists from diverse backgrounds are less likely to obtain grant funding, publish as first author, and get promoted (Stevens et al., 2021).

Research teams emphasized that efforts to be intentional and plan ahead to prepare for additional costs related to this work are undermined by budget constraints. Funding agencies, as well as those responsible for approving proposals and distributing budgets, should be required to gain competencies in the challenges and costs associated with nontraditional research approaches to enhance inclusion.

Optimizing the Study Team to Ensure Alignment with Research Goals

All of the research teams that were interviewed described the composition of the study team as an important component of representative research. Research teams interact with potential study participants and are instrumental in the success of recruitment and retention. Diverse study teams were generally described as being helpful to recruitment to enhance congruence between research teams and potential participants, and this congruence was described in different ways depending upon the focus of the study (e.g., age, sex, race, ethnicity). Retaining study staff over time was also emphasized as very important to recruitment and retention success; however, this may be difficult given the competition for skilled study staff. For example, "So having the same staff at our site, we've had the same staff for 11 years now and are so thankful and grateful. And we've done everything to retain the staff . . . because they're the face of the study." It is important to note that cultural and linguistic congruence with the target population was not enough. Gaining engagement and community buy-in for the study goals and desired outcomes were equally important when working with communities that are underrepresented in clinical trials and research. Several studies show that increasing the diversity of study staff and leadership leads to increased enrollment of diverse populations and improved reporting of results (Khan et al., 2020; Nielsen et al., 2017; Whitelaw et al., 2021). One study found that there was a greater likelihood of reporting sex-stratified results when a woman was either first or last author (Nielsen et al., 2017). Additionally, studies have demonstrated a positive association between the number of women as coauthors and a higher proportion of women participants in the research (Reza et al., 2020).

More can be done to train and develop the next generation of diverse principal investigators. Academic research institutions play a key role in diversifying principal investigators as they train a large percentage of the research workforce, including investigators and research staff. They have the opportunity to diversify the pathway of the students and future clinician-scientists entering health science professions and Ph.D. programs (see Box 5-1). To increase recruitment, retention, and advancement of diverse faculty, institutions can follow and invest in evidence-based practices, as described in Box 5-2.

Academic medical centers also play an important role in investing in and supporting research that designs and tests new strategies that are practical and pragmatic to enhance diverse recruitment of participants representative of the population with a given disease. Academic research institutions can offer training on systemic racism in research, implicit bias, and cultural sensitivities to researchers and research staff. They can also educate researchers on strategies to increase diverse enrollment, including use of broad eligibility criteria and avoiding sex-specific exclusion criteria.

In addition to efforts by academic medical centers, professional societies and federal agencies also have influence by providing training programs for both early- and mid-career women and underrepresented scientists. One example is the American College of Cardiology's Clinical Trials Research: Upping Your Game program, which is designed to train the next generation of clinical trial team scientists by developing women and underrepresented populations in cardiology. This program includes three 2-day sessions that focus on clinical trials research; networking with other clinical trialists, investigators, industry leaders, and regulatory stakeholders; and developing a personal career action plan (ACC, 2022). Another example is the NIH's Faculty Institutional Recruitment for Sustainable Transformation (FIRST) program, which aims to enhance and maintain cultures of inclusive excellence in the biomedical research community. These FIRST awards are given to academic institutions to recruit cohorts of early-career faculty who are competitive for assistant professor positions and have demonstrated commitment to inclusive excellence (NIH, 2021c). This is a relatively new program, so evaluation of the program is not available yet. However, the committee feels that these types of initiatives that encourage and promote enhancing diversity and inclusion in institutional contexts are critical for developing our future workforce.

> **BOX 5-1**
> **Federal Support of Early-Career Researchers**
> **Can Affect Access to Opportunities**
>
> The nature of federal research to support early-career professionals in a biomedical career receive can affect the extent to which they gain access to mentorship and professional development opportunities that can make a difference in supporting their career growth and advancement. In 2016, only about 10 percent of postdoctoral researchers in the biomedical, behavioral, social, and clinical sciences were supported on federal fellowships and traineeships, such as the Ruth L. Kirschstein Individual National Research Service Award (NRSA) postdoctoral fellowship (F32), which provides support to individual postdoctoral fellows, and the Ruth L. Kirschstein NRSA Institutional Research Training Grant (T32), which provides support to institutions to develop training opportunities for selected individuals. These fellowships and traineeships undergo peer review of the research and training plan, and include stipulations for professional development and mentoring by eligible mentors. In contrast to the fellowships and traineeships, other mechanisms of early-career support do not generally include as a requirement a plan for training and professional development in the grant application, an assessment of the principal investigator as a mentor, nor any other formal mechanism to ensure quality training and mentorship opportunities.
>
> Furthermore, National Institutes of Health (NIH) training grants, such as individual F32 and institutional T32 awards, are restricted to U.S. citizens and permanent residents, yet a substantial proportion of biomedical postdoctoral researchers are not U.S. citizens or permanent residents. Of the biomedical postdoctoral researchers included in the 2015 National Science Foundation General Social Survey, 53 percent held temporary visas and 31 percent reported earning their degrees in a foreign country. These percentages reflect both the openness of U.S. biomedical training and labor markets and the attractiveness of U.S. research careers to international scholars. Some countries also encourage recent Ph.Ds to seek postdoctoral training in the United States. Lack of access to the F32 and T32 programs may undermine advancement of talented individuals who are contributing meaningfully to U.S. research and medicine on the basis of citizenship status. In addition, the National Institute of General Medical Sciences now requires T32 applicants to submit a Recruitment Plan to Enhance Diversity and Trainee Retention Plans. This could be expanded across all NIH institutes to ensure that training grants are supporting a diverse biomedical workforce (NIGMS, 2021).
>
> SOURCE: Content adapted from NASEM, 2018.

BOX 5-2
Promising Practices for Supporting a More Diverse and Equitable Medical Workforce

A growing body of research literature and an increasing number of examples identify strategies and practices that institutions and organizations can adopt to diversify talent pools, mitigate biases in evaluation and promotion, and create and sustain a positive, inclusive organizational climate. Among those practices, organizations should consider adopting and adapting—in concert with evaluation to understand the impact of these interventions on their communities and within their institutional context—are the following:

To Recruit a Diverse Applicant Pool:

- Work continuously to identify promising candidates from underrepresented groups and expand the networks from which candidates are drawn.
- Write job advertisements that appeal to a broad applicant pool and use a range of media outlets and forms to advertise these opportunities broadly.
- Eliminate or lessen the emphasis given to admissions requirements that are particularly subject to bias or may be poor predictors of success (e.g., certain standardized test scores).
- Decide on the relative weight and priority of different admissions or employment criteria before interviewing candidates or applicants.
- Hold those responsible for admissions and hiring decisions accountable for outcomes at every stage of the application and selection process.
- Educate evaluators to be mindful of the childcare and family leave responsibilities often faced by women, especially when considering "gaps" in a resume.
- When possible, use structured interviews in admission and hiring decisions.

To Improve Retention:

- Ensure fair and equitable access to resources for all employees and students.
- Broadly communicate about the institutional resources that are available to students and employees and be transparent about how these resources are allocated.
- Set and widely share standards of behavior, including sanctions for disrespect, incivility, and harassment.
- Create and widely advertise policies and practices that address workers' need to balance work and family roles throughout their education or careers.
- Support mentorship initiatives that recognize, respond to, value, and build upon the power of diversity.
- Create "counterspaces" that provide a sense of belonging and support and serve as havens from isolation and microaggressions.

continued

> **BOX 5-2 Continued**
>
> **To Improve Advancement:**
>
> - Create sponsorship programs through which individuals with positions of power and influence advocate publicly for the advancement of talented individuals to senior leadership positions.
> - Establish clear metrics for success and advancement and avoid reliance on metrics that are known to be biased (e.g., teaching evaluations, impact factor of publications, appraisal of "potential").
> - Mitigate bias in performance evaluations, promotion decisions, and selections for awards and special recognitions.
>
> SOURCE: Content adapted from NASEM, 2020a.

Attaining Resources and Support to Achieve Representativeness

The investment of time and money are necessary to successfully engage in the long-term strategies and relationship building needed to drive inclusion in studies (Green-Harris et al., 2019). According to the research teams that were interviewed, funding for these recruitment efforts was of paramount importance, requiring special funding announcements focused on inclusion of underrepresented groups, expanded budgets for teams attempting to recruit and retain these groups, and flexibility within budgets to allow for deeper engagement of community partners. For example,

> I think that it would be good for efforts to recruit and retain these folks, to have potential additional budgeting so like it's a $500,000 grant but you're going to recruit over 40 percent folks with lower socioeconomic status, then there's an extra $50,000 a year for direct costs to support those efforts. I think we have to put our money where our mouth is, and I don't see that is happening.

In addition to funding, research teams emphasized education of researchers and providing supports such as professional networks and institutional resources with expertise in these areas. Finally, material support for community organizations so that they can build infrastructure to enhance enrollment in clinical studies also emerged as an important long-term necessity to enhance inclusion. In particular, resources that could assist these organizations in building an ongoing foundation for research would create successful long-term partnerships (George et al., 2014). Investments in community-based strategies and partnerships are needed to help minimize the power imbalance between the researcher and participant in ways that build trust in research teams and institutions (BeLue et al., 2006).

The need for greater investments in the people, communities, and institutions engaged in research is echoed in Michos et al. (2021), which outlines several

FACILITATORS OF SUCCESSFUL INCLUSION IN CLINICAL RESEARCH 119

large- and small-scale interventions by stakeholders for improving enrollment and reflecting the diverse U.S. population in research (see Figure 5-1) (see also Michos and Van Spall, 2021).

Investing in community-based research is critical for developing relationships and involving communities in clinical research. Community-based research takes place where people live, work, and play. Effective community-based research settings create a bridge between the community, scientific institutions, and researchers and build trusting partnerships that are essential for successful research participation. Together researchers and community members engage in the design and conduct of research with the goal of building trust and respect for the values, viewpoints, and interests of the community members. Specific examples of these partnerships are described in the Academic Institutions section, below.

There are many ways to involve community members in ongoing research. UCSF Accelerate (2022) provides a step-by-step guide for practicing community-engaged research, such as the following:

- Assemble a research team that includes community clinicians, clinic staff, and community members who are decision-makers. In addition, set up a patient advisory board that is involved throughout the process.
- In coordination with community clinicians and advisors, identify issues of greatest need and importance to ensure research is relevant and resonates with the community.

FIGURE 5-1 Improving diversity in enrollment.
NOTE: Although the right side of the figure reads "diverse populations," in the context of this report, the committee is using this figure to specifically improve enrollment of underrepresented populations in trials.
SOURCE: Reprinted with permission from Michos et al., 2021.

- Involve community clinicians and advisors in the writing process and determining study questions to address.
- Communicate the relevance of the study design, but also be prepared to modify the design with more community acceptable approaches, which may involve gathering focus groups or other qualitative measurements.
- Review findings with community members and disseminate results in a way that is appropriate to the community members.
- Include community clinicians and advisors as authors on scientific papers and presenters in community and broader settings.

Investing in the science of engagement and empowerment can also help overcome barriers to equity. As described below, funding agencies, institutions, and researchers all have a role to play in improving community engagement and empowerment.

Funding Agencies

Major funders have a mandate (and/or vested interest) to demonstrate return on investment. For example, the NIH budget is established and renewed by the U.S. Congress, which is responsible for its oversight. Federal research awards are typically funded for a period ranging from 2 to 5 years, a time frame that is meant to encompass all phases of research from project startup to results dissemination. Funding periods and budgets often discourage researchers from more participatory and emancipatory methods. Some funders are moving away from these models, and folding in stakeholder engagement as a major requirement of funding. For example, the Patient-Centered Outcomes Research Institute has invested nearly $3 billion in comparative effectiveness research on health since FY 2010 (PCORI, 2020). The institute engages patients and providers in identifying research priorities, trains patients to review and evaluate applications for funding, and requires that patient engagement be documented in every step of a research project, from the formulation of the research question to the research methods to dissemination of results (see Box 5-3).

Academic Institutions

Academic institutions play a significant—if sometimes obscured—role in community empowerment and engagement (as described in Chapter 4). For example, institutions set expectations for faculty productivity, which have impacts on the extent to which their faculty invest in community-engaged research. As noted above, academic institutions play a critical role in recruiting and retaining diverse faculty and investing in the future workforce.

Institutions and their surrounding communities also have natural ties, but these ties have not always benefited community members. Attention to issues

BOX 5-3
Patient-Centered Outcomes Research Institute: Supporting Engagement

The Patient-Centered Outcomes Research Institute's (PCORI) mission is to advance patient-centered, stakeholder-engaged research throughout the research process. PCORI engagement principles include approaches to integrate equity and inclusion across the research enterprise. PCORI provides funding support, tools, and resources to research stakeholders.

The Eugene Washington PCORI Engagement Awards Program provides funding for research support projects that encourage involvement of patients, caregivers, clinicians, and other health-care stakeholders as integral members of the patient-centered outcomes research/comparative clinical effectiveness research (CER) enterprise. PCORI funding opportunities include awards for three types of engagement projects:

- Capacity Building: Projects that help communities increase their facility with and ability to participate across all phases of the PCOR/CER process.
- Dissemination Initiative: Projects that help organizations and communities plan for or actively bring relevant PCORI-funded research findings to end users and encourage use of this information in their health-care decision making.
- Stakeholder Convening Support: Projects that include multistakeholder convenings, meetings, and conferences that align with PCORI's mission and facilitate expansion of patient-centered outcomes research/comparative clinical effectiveness research, or PCOR/CER, through collaboration on such efforts.

In fiscal year 2022, PCORI plans to award up to $25 million as part of the Eugene Washington PCORI Engagement Awards Program.[a]

To enhance the uptake of engagement practices and methodologies within the broader health-care research community, PCORI maintains a repository of tools and resources, known as the Engagement Tool and Resource Repository for Patient-Centered Outcomes Research. This repository is focused on research engagement and capacity building across the project lifespan and is searchable by focus area, health condition, stakeholder audience, targeted population, and phase of research in which the engagement occurred.

PCORI also provides a variety of engagement information to support research stakeholders. For example, PCORI offers resources for building and supporting effective multistakeholder research teams, methods for engaging stakeholder partners throughout a research study, and guidance to help researchers identify budgetary items associated with engagement within a research study.[b]

---

[a] See https://www.pcori.org/engagement/eugene-washington-pcori-engagement-awards.
[b] For more information and additional resources, see https://www.pcori.org/engagement/engagement-resources.

such as gentrification, local needs, areas of mutual interest, and sustainability can foster engagement and empowerment with community members. Institutions also have the flexibility to leverage internal funds (revenue, endowments) for community engagement and development. For example, in 2015 Indiana University announced it would dedicate $300 million to address health issues important to Indiana communities. The Grand Challenges program funded three major projects with distinct health foci, including precision health, environmental resiliency, and substance use disorders.[2] While these efforts are currently under way, they have led to an expansion of federal funds. The Grand Challenge on substance use disorder, for example, has helped hundreds of Indiana teens involved in the criminal justice system get screened for substance use issues, and has now expanded to eight additional counties with the help of a recent grant from the National Institute on Drug Abuse.[3] The Wisconsin Alzheimer's Institute, Regional Milwaukee Office, is another example of an institution investing in communities to address issues and build relationships with great success (see Box 5-4).

Many institutions also have affiliated health centers, which can play a key role in community engagement and investment. Many of these health centers partnered with communities to rapidly react to the outbreak of the COVID-19 pandemic, showing that this model is possible and effective for the health of the public. For example, the University of New Mexico partnered with the city of Albuquerque, local health departments, nonprofits, and more to assist seniors and individuals struggling with homelessness during the COVID-19 pandemic.[4] In another example, the NIH-funded California Community Engagement Alliance (CEAL) consortium of 11 community-academic teams across the state (including academic health centers, community clinics, community-based organizations) developed locally tailored strategies to promote effective communication about COVID-19, improve participation of underrepresented groups in vaccine and therapeutic research, increase vaccine uptake, and enhance clinical and public health equity for the communities hardest hit by the pandemic (AuYoung et al., 2022; Stadnick et al., 2022).

CONCLUSIONS

This chapter provides evidence-based key themes that emerged from a qualitative study of 20 study investigators and staff to promote representation in clinical studies, and it delineates practical and innovative approaches for various stakeholders involved in the clinical research enterprise, including principal investigators, research staff, academic institutions and the broader scientific community, community-based organizations, community clinics, public health

---

[2] See https://grandchallenges.iu.edu/.
[3] See https://addictions.iu.edu/news/recovery-month-2020.html.
[4] A full review of these partnerships can be found at https://www.ncbi.nlm.nih.gov/pmc/articles/PMC7380298/.

BOX 5-4
Case Study: The Wisconsin Alzheimer's Institute, Regional Milwaukee Office Community Engagement

The Wisconsin Alzheimer's Institute, Regional Milwaukee Office (WAI Milwaukee), with the support of Bader Philanthropies and the University of Wisconsin School of Medicine and Public Health, takes an exemplary approach to community engagement and empowerment using an asset-based community development approach. Since 2008, the WAI Milwaukee program has worked closely with the Milwaukee and Southeastern Wisconsin African American community to improve the diagnosis of Alzheimer's disease in its aging population. It works with community leaders, partners, and stakeholders to provide appropriate and culturally specific health-care and supportive services, raise awareness of Alzheimer's disease, and increase participation in research. Research is among the program's priorities, but supporting the community, while having an impact on the systemic causes of disparities and systemic barriers, is the primary goal.

The WAI Milwaukee program fosters community empowerment to address health disparities and the lack of participation in research. An emphasis on the community's strengths is the focus of these five integrated mission areas of the WAI Milwaukee program: Community Engagement, Community and Professional Education, Service, Advocacy, and Research.

Investment in the community by the WAI Milwaukee program has led to successful engagement, involvement, and commitment from the community. Investment is not simply giving information to the community; the WAI Milwaukee program devotes substantial time and resources fostering relationships with the community. Building relationships and trust, while acknowledging the health needs of the community, are the foundation of the program's community engagement activities, and these are prioritized well above a focus on the scientific needs and research participation.

organizations, recruitment centers, pharmaceutical companies, professional organizations, funding agencies, institutional review boards, and journals. Ultimately, efforts to improve representation should involve provision of financial resources for research teams, long-term infrastructure based in communities, material and social support for community advocates and organizations, and education about the relevance of these efforts to scientists, community members, and relevant stakeholders, as well as potential study participants and their caregivers. Dedicated and ongoing funding will be essential to build the infrastructure to achieve representation, and community stakeholders will need to be included and engaged at every step to achieve these goals.

TABLE 5-1 Strategies to Achieve Representation in Clinical Research by Theme

Starting with intention and agency to achieve representativeness

- Budget for time, staff, and resources needed to conduct this work in an ethical and equitable manner.
- Value the work to elucidate pathways of diseases or mechanisms of action of interventions in underrepresented groups.
- Approach the work with persons from underrepresented communities with a sense of ethical and fiduciary responsibilities.
- Highlight the benefits of research to provide access to innovative interventions that may otherwise not be available.

Establishing a foundation of trust with participants and community

- Acknowledge the abuse, both historical and current, that many underrepresented groups have experienced in research.
- Recognize that trust is fragile.
- Incorporate community advisory boards as equitable partners in research.
- Participate in community outreach through educational events, health fairs, and other venues.
- Develop lasting relationships with study participants through regular contact and updates on the study.
- Avoid conducting "helicopter" research by incorporating periods before study recruitment to build community relationships.
- Bring research to the community in the places where community members live, work, and play.
- Create personal connections with each participant using an individualized approach that is genuine.
- Listen to community members and incorporate their needs into future research agendas and subsequent projects.
- Provide incentives to caregivers and/or identify aspects of the protocol that can be provided as free services to persons accompanying participants to visits.
- Develop study materials that are appropriate for the patient's literacy level and linguistic background.

TABLE 5-1 Continued

Streamlining enrollment criteria to promote inclusivity without compromising scientific rigor.

- Reduce the burden of participation by offering alternatives to in-person visits to the research center (home visits, remote visits).
- Incorporate technology to streamline processes (e.g., online consenting) and training and support for that technology, if needed.
- Engage cultural experts to assist with developing culturally sensitive study protocols.
- Make research teams accessible via several platforms (e.g., website, email, landline).
- Recognize the heterogeneity of participants and adjust recruitment approaches accordingly.

Adopting a flexible approach to recruitment and data collection

- Adapt study protocols throughout the study in response to participant feedback.
- Tailor outreach efforts to the participant's needs and seek community representatives to assist with tailoring these efforts.
- Institute buddy systems where participants are allowed to share rides or complete aspects of the study on the same day.
- Allow for partial completion of visits.
- Seek and adopt feedback from community members when protocols are not yielding results.

Building a robust network by identifying all relevant stakeholders

- Incorporate community advisory boards as equitable partners in research.
- Elicit perspectives of frontline staff and potential participants to optimize study protocols and community engagement.
- Identify elements of the protocol that could benefit caregivers and provide incentives to engaging caregivers.

Navigating scientific, professional peer, and social expectations

- Increase representativeness of professionals from diverse communities into decision-making positions (e.g., review panels, journal editors).
- Learn about principles of community-based participatory research.
- Invite scientists and study staff to observe existing efforts by successful groups in engaging diverse communities.
- Create networks for scientists focused on recruitment and retention of certain groups.

Optimizing the study team to ensure alignment with research goals

- Hire and retain diverse and experienced staff members.
- Provide training for staff in the form of observation and regular team meetings.
- Provide training in implicit bias and strategies to address its effects on interactions with participants and across the research team.
- Seek out staff members who are committed to the cause of the study team and clinical research in general.
- Strive to engage all members of the community, even if not the population affected by the condition of study (e.g., organize educational workshops about healthy eating for everyone, even if studying cardiovascular health in older adults).
- Recruit members of the target community, and others with lived experience, as study team members.

continued

TABLE 5-1 Continued

Attaining resources and support to achieve representativeness
- Create funding announcements to support inclusion of diverse groups in research studies.
- Include community partners as sites on a grant submission. Be mindful of hierarchical approach with academic institution as lead and strive to create more equitable collaborations.
- Allow for flexibility in use of funding to incentivize clinicians, administrators, and stakeholders providing research support.
- Provide reviewer training/instructions on diversity in recruitment.
- Create new funding mechanisms with fewer constraints on budget and time frame than existing mechanisms (e.g., R01).
- Ensure that resources material/knowledge/skills endure in the community.
- Develop partnerships with community leaders and members so that researchers can leverage these resources.

6

Recommendations for Improving Representation in Clinical Trials and Clinical Research

CONCLUSIONS

1. **Improving representation in clinical research is urgent.**

    The scientific necessity to improve research equity is urgent. The United States is becoming more diverse, with the 2020 U.S. Census finding that the number of people who identify as white has decreased for the first time since a census started being taken in 1790. Despite greater diversity, deep disparities in health are persistent, pervasive, and costly. Without major advancements in the inclusion of underrepresented and excluded populations in health research, meaningful reductions in disparities in chronic diseases such as diabetes, cancer, and Alzheimer's disease remain unlikely. Purposeful and deliberate change is needed. As the United States becomes more diverse every day, failing to reach these growing communities will only prove more costly over time (see Chapter 2).

2. **Improving representation in clinical research requires investment.**

    Improving the representation of underrepresented and excluded populations in clinical trials and clinical research requires a substantial investment of time, money, and effort. Investment of time and resources are needed to build and restore trust with underrepresented and excluded communities. Building trust with local communities cannot be episodic or transactional and pursued only to meet the goals of specific studies; it requires sustained presence, commitment, and investment. Investments are also needed in the systems and technologies that reduce burdens to participation by underrepresented and excluded populations, such as by adequately compensating participants financially for their time when participating in research and by investing resources in making participation more physically accessible. Lastly, we need to invest in creating a more diverse workforce that better reflects the diversity of our country. This not only has implications for study-site personnel and their direct interactions with participants, but also influences the types of research questions that get asked, the types of research that gets funded, and even the types of research that are published. To better address health disparities and ensure health equity for all, the U.S. workforce should look more like the nation (see Chapter 4).

3. **Improving representation requires transparency and accountability.**

Transparency and accountability throughout the entire research enterprise will be critical to driving change and must be present at all points in the research life cycle—from the questions being addressed, to ensuring the populations most affected by the health problems are engaged and considered in the design of the study, to recruitment and retention of study participants, to analysis and reporting of results. Individual investigators and research institutions on the front lines bear responsibility for transparency in reporting progress toward the goals of inclusion in research, but this must be reinforced by transparency and accountability that funding agencies and industry sponsors have across their portfolios, that regulatory agencies have in their role governing the conduct of research as well as the approval and reimbursement of the drugs and devices that are often the final products of clinical research, and that journal editors and others that disseminate research have in communicating findings (see Chapters 3, 4, and 5).

4. **Improving representation in clinical research is the responsibility of everyone involved in the clinical research enterprise.**

The clinical research landscape is complex and involves multiple stakeholders—participants, communities, investigators, institutional review boards, industry sponsors, institutions, funders, regulators, journals, and policy makers. Each of these stakeholders has a critical role to play in achieving the goal of improving representation in clinical research, but the complex nature of the research ecosystem and research processes, combined with lack of accountability and historic underinvestment means that an issue that should be everyone's responsibility can become no one's priority. In this report, the committee emphasizes that the research supports taking a systematic approach to addressing this issue; one in which all stakeholders take responsibility for the important role they can play in supporting representation in clinical research participation.

The committee was asked, "Who bears the cost of more inclusive science?" The responsibility (and therefore the cost) will be borne to some extent by all stakeholders in the larger research ecosystem, acting in concert to achieve this larger societal and scientific goal. Those that profit from scientific discovery bear particular responsibility in shouldering the cost of inclusivity. The federal government has a notably prominent role and responsibility in achieving the goal of more inclusive research, as a primary funder of the research enterprise with taxpayer dollars, regulator of the processes of scientific research, gatekeeper to approvals for monetizing scientific discovery, and purchaser of new drugs and devices. More coherence of federal policy to align investment and accountability to achieve the goals of inclusive science is warranted.

In answering the question of who bears the cost of more inclusive science, we must also ask, "Who bears the cost of the current lack of inclusivity?" That cost is large (as evidenced by the analysis in Chapter 2), is borne disproportionately by underrepresented and historically excluded communities, but saps the health and economic strength of the entire society.

5. **Creating a more equitable future entails a paradigm shift.**

The committee sees the need for both pragmatic approaches and an aspirational vision. To realize a more equitable future, the report epilogue implores the field to embrace a paradigm shift that moves the balance of power from institutions and puts at the center the priorities, interests, and voices of the community. An ideal clinical trial and clinical research enterprise pursues justice in the science of inclusion through scalable frameworks; expects transparency and accountability; invests more in people, institutions, and communities to drive equity; and invests in the science of community engagement and empowerment. These ideals should be the foundation of the actions that stakeholders take to make sustainable change.

RECOMMENDATIONS

The committee's recommendations focus on tangible actions that must urgently be taken within the context of the existing structures of the clinical research ecosystem in order to achieve the goals of representation and inclusion. Although individual researchers can take many actions to improve equity in clinical trials and clinical research, as described in Chapter 5, the committee focused on system-level recommendations to drive change on a broader scale. The committee presents 17 recommendations to improve the representation of underrepresented and excluded populations in clinical trials and clinical research and create lasting change.

The urgency of addressing the equity in research participation and the lack of substantial progress despite stated commitments led the committee to propose bold recommendations with potentially far-reaching implications. The committee is aware that the complexity of the U.S. health-care system poses significant challenges to transforming the clinical research system, and these systematic challenges will also influence the implementation of the committee's recommendations. While providing a complete policy assessment for each recommendation was outside of the committee's scope and charge, the committee does not deny that there will be costs—both fiscal and political—associated with the implementation of the recommendations. These costs must be carefully weighed against the potential for long-term benefit. Changing our nation's approach to clinical research may require significant upfront costs to more equitably recruit and retain a diverse group of participants and to hold investigators accountable when they do not meet these goals. In addition, it will require incentivizing sponsors of clinical research to change the status quo. However, based on the committee's expert opinion and the available evidence, the committee believes that implementation of its recommendations is necessary to truly drive significant and sustained change to the clinical research system.

Reporting and Accountability

1. **The Department of Health and Human Services (HHS) should establish an intradepartmental task force on research equity charged with coordinating data collection and developing better accrual tracking systems across federal agencies, including the Food and Drug Administration (FDA), National Institutes of Health (NIH), Centers for Disease Control and Prevention (CDC), Agency for Healthcare Research and Quality (AHRQ), Health Resources Services Administration (HRSA), Indian Health Services (IHS), Centers for Medicare & Medicaid Services (CMS), and two departments outside the Department of HHS, the Department of Veterans Affairs and Department of Defense. This task force should be charged with the following:**
   a. Producing an annual report to Congress on the status of clinical trial and clinical research enrollment in the United States, including the number of patients recruited into clinical studies by phase and condition; their age, sex, gender, race, ethnicity, and trial location (i.e., where participants are recruited); their representativeness of the conditions under investigation; and the research sponsors.
   b. Making data more accessible and transparent throughout the year, such as through a data dashboard that is updated in real time.
   c. Determining what "representativeness" means for protocols and product development plans.
   d. Developing explicit guidance on equitable compensation to research participants and their caregivers, including differential compensation for those who will bear a financial burden to participate.
2. **The FDA should require study sponsors to submit a detailed recruitment plan no later than at the time of Investigational New Drug and Investigational Device Exemption application submission that explains how they will ensure that the trial population appropriately**

reflects the demographics of the disease or condition under study and that provides a justification if these enrollment targets do not match the demographics of the intended patient population in the United States.
3. The NIH should standardize the submission of demographic characteristics for trials to ClinicalTrials.gov beyond existing guidelines so that trial characteristics are labeled uniformly across the database and can be easily disaggregated, exported, and analyzed by the public. The data reported should include the number of patients; their age, sex, gender, race, ethnicity, and trial location (i.e., where participants are recruited); who sponsors them; and language accessibility.
4. In grant proposal review, the NIH should formally incorporate considerations of participant representativeness in the score-driving criteria that assess the scientific integrity and overall impact of a grant proposal. These criteria should be part of the assessment of the scientific approach, including whether it is appropriate for generating insights for the populations to whom the results are intended to generalize. The criteria should also be incorporated in the assessment of whether investigative teams and environment have detailed and feasible plans to meet the goals of representative study enrollment. Additionally, the NIH should assess in its annual review of progress reports of funded studies whether a given study has met the proposed enrollment goals of representativeness by race/ethnicity, sex, and gender, and should establish a plan for remediation for the investigator and/or organization that includes criteria for putting funding on hold that has not met predefined recruitment goals.
5. Journal editors, publishers, and the International Committee on Medical Journal Editors should require information on the representativeness of trials and studies for submissions to their journals, particularly relative to the affected population; should consider this information in accepting submissions; and should publish this information for accepted manuscripts. The information required should include the following:
   a. The disease, problem, or condition under investigation
   b. Special considerations related to sex and gender, age, race or ethnic group, and geography
   c. The overall representativeness of the trial, including how well the study population aligns with the target population in which the results are intended to generalize. If the study population does not align with the population affected by the disease, authors should provide scientific justification for why this is the case.
6. The Office of Human Research Protections (OHRP) and the FDA should direct local institutional review boards (IRBs) to assess and re-

port the representativeness of clinical trials as one measure of sound research design that it requires for the protection of human subjects. Representativeness should be measured by comparing planned trial enrollment to disease prevalence by sex, age, race, and ethnicity in the trial location (i.e., where participants are recruited). Protocols in which the planned enrollment diverges substantially from disease prevalence should require justification. The OHRP and FDA should establish a plan for remediation for local IRBs that frequently approve protocols that are not representative.

7. The CMS should amend its guidance for coverage with evidence development to require that study protocols include the following:
   a. A plan for recruiting and retaining participants that are representative of the affected beneficiary population in age, race, ethnicity, sex, and gender.
   b. A plan for monitoring achievement of representativeness as described above, and a process for remediation if CED studies are not meeting goals for representativeness.

Federal Incentives

8. In order to determine how to take action on the most effective accountability and incentive structures, Congress should direct the FDA to enforce existing accountability measures, as well as establish a taskforce to study new incentives for new drug and device applications for trials that achieve representative enrollment. Incentive programs should be designed to improve representativeness in clinical research, improve clinical outcomes, and ensure they do not reduce access to new therapies. Some ideas include:
   a. Tax incentives, such as tax credits for research and development
   b. Fast-Track criteria and exemption from some FDA drug application fees
   c. Extended market exclusivity to sponsors who meet predefined criteria of representativeness
   d. Refusing to file an application that does not appropriately represent the target population under study
9. The CMS should expedite coverage decisions for drugs and devices that have been approved based on clinical development programs that are representative of the populations most affected by the treatable condition.
10. The CMS should incentivize community providers to enroll and retain participants in clinical trials by reimbursing for the time and infrastructure that is required. Through the creation of new payment codes, CMS should reimburse activities associated with clinical trial

participation, including but not limited to data collection and personnel (e.g., community health workers, patient navigators) to support research education and recruitment.
11. The Government Accountability Office (GAO) should assess the impact of reimbursing routine care costs associated with clinical trial participation for both Medicare (enacted in 2000) and Medicaid (enacted in 2020). The assessment should include an analysis of whether there is timely and complete reimbursement, any implications for innovation and care delivery to underrepresented populations, and any challenges to implementation.

Remuneration

12. Federal regulatory agencies, including the OHRP, NIH, and FDA, should develop explicit guidance to direct local IRBs on equitable compensation to research participants and their caregivers. In recognition that research participation may pose greater hardship or burdens for historically underrepresented groups, the new guidance should encourage and allow for differential compensation to research participants and their caregivers according to the time and financial burdens of their participation. Differential compensation may include additional reimbursement for expenses including but not limited to lost wages for those with lower socioeconomic status (SES), transportation costs, per diem, dependent care, and housing/lodging where applicable.
13. All sponsors of clinical trials and clinical research (e.g., federal, foundation, private and/or industry) should ensure that trials provide adequate compensation for research participants. This compensation may include additional reimbursement for expenses including but not limited to lost wages for lower SES participants and family caregivers, transportation costs, per diem, dependent care, and housing/lodging where applicable.

Education, Workforce, and Partnerships

14. All entities involved in the conduct of clinical trials and clinical research (academic centers, health-care systems, sponsors, regulatory agencies, and industry) should ensure a diverse and inclusive workforce, especially in leadership positions.
15. Leaders and faculty of academic medical centers and large health systems should recognize research and professional efforts to advance community-engaged scholarship and other research to enhance the representativeness of clinical trials as areas of excellence for promotion or tenure.

16. Leaders of academic medical centers and large health systems should provide training in community engagement and in principles of diversity, equity, and inclusion for all study investigators, research grants administration, and IRB staff as a part of the required training for any persons engaging in research involving human subjects. This training should incorporate strategies to enhance diverse recruitment and retention in clinical research, as well as planning of and budgeting for these efforts and timely reimbursement of partnering agencies and organizations.
17. HHS should substantially invest in community research infrastructure that will improve representation in clinical trials and clinical research. This funding should go to agencies such as the HRSA, NIH, AHRQ, CDC, and IHS to expand the capacity of community health centers and safety net hospitals to participate in and initiate clinical research focused on conditions that disproportionately affect the patient populations they serve.

Epilogue: Envisioning a New Future

The findings, conclusions, and recommendations in this report make clear the implications of maintaining the status quo and the critical need to find new ways to achieve greater representation in clinical research. The report provides evidence-based recommendations, which, when implemented, would move the nation closer to a more equitable and just society. However, the committee would also like to acknowledge that all of the evidence cited in this report is derivative of a system that is fundamentally oppressive and problematic. Thus, our evidence-based recommendations are constricted by the same forces. This epilogue summarizes some of the large, system-level changes that the committee would like to see in order to truly realize an inclusive and representative clinical research landscape in the United States that leads to greater justice and health equity in this nation.

The committee believes that to improve representativeness in research effectively and sustainably, progress must be made in both the development of a rigorous science of inclusion and in the pursuit of theoretical frameworks that investigate and challenge the "socio-political determinants of exclusion" (Gilmore-Bykovskyi et al., 2021). This combined effort can help deliver a needed paradigm shift in the balance of power from institutions to communities.

THE SCIENCE OF INCLUSION

First, the committee believes there must be intentional efforts to support the development of a rigorous science of inclusion and community engagement. Health equity scholars have long highlighted the importance of investing in the development and adoption of evidenced-based strategies to move the field toward scalable frameworks for engagement, recruitment, and retention (Curry and Jackson, 2003; Dilworth-Anderson, 2011). However, research in this area often evaluates individual-level beliefs and attitudes, site-specific barriers, and qualitative approaches to understanding trust, which are constraints on the evidence presented in earlier chapters of this report. Methodically rigorous approaches are needed to move beyond individual person- and site-specific barriers to facilitate system-level change. This requires a focus and investment in the development of interdisciplinary teams that include community representatives to develop rigorous empiric evaluations of strategies and approaches for driving inclusion in research. As described in Chapter 2, improving these systems is not just desirable, but necessary for a more just, healthy, and equitable world.

Defining Inclusion

Second, the committee recognizes that this report represents one step on the path toward inclusion. Throughout this report, the committee has used the phrase underrepresented and excluded populations, but have focused specifically on women and racial and ethnic minority population groups, as defined by the committee's charge. As a result, the committee has not focused on the unique needs of rural, frontier, transgender, non-binary, neurodivergent, disabled, lower socio-economic status, illiterate, elderly, pediatric, and countless other populations. The needs of these populations and their contributions to research are as critically important, and no less urgent, than the populations highlighted by the committee in this report. The recognition and inclusion of all underrepresented and excluded populations is an urgent problem that needs to be addressed. It is critical that the clinical trials and research community examine how underrepresented and excluded populations are defined, and who is included or excluded by that definition. The clinical trials and research community will continue to work within a narrow definition of inclusion unless action is taken to change it. This is why the structural changes, at all levels and across systems, recommended by this report are so urgently needed. This is also why it is critical to center whole communities, and not simply specific communities, as part of the research process.

EMBRACING JUSTICE

Third, realizing this vision will require not only the rethinking of conventional practices and investments as outlined in the report but also in the adoption of new theoretical frameworks for conceptualizing research centered in equity and social justice. This includes a close interrogation and understanding of the factors that contribute to the status quo. In the literature, these factors are often presented at the individual level, such as participant trust and beliefs about research, religiosity, and willingness. However, deep gaps still exist in the understanding of the problem, especially when the onus of responsibility for improving engagement and participation is placed on the individual participant, rather than on institutions and researchers. The committee believes there is a need for stakeholders to broaden the possibilities for transformative solutions by bringing into the forefront the historical, institutional, and social contexts that shape research accessibility. This more emancipatory approach can encourage research stakeholders to reflect and act on the injustice that exists in the communities where they work, enabling actions that center communities and advance justice in the research process (Wesp et al., 2018). It moves beyond equitable engagement of communities in research questions, studies, and processes that already exist to transforming the research enterprise for science beneficial to those communities (Wilkins and Alberti, 2019). While this approach is challenging, the report offers examples of this approach working. For example, Box 5-4 describes the Wisconsin Alzheimer's Institute Regional Milwaukee Office and its ongoing efforts to invest in community empowerment to address health disparities. Examples like this provide evidence that a paradigm shift is possible and can help advance community health through more equitable research practices.

According to Gilmore-Bykovskyi et al. (2021), "Fulfilling justice in research is foundational to cultivating practices that promote health equity through equal valuation of the wellbeing of all persons, the correction of injustices, and providing resources according to need, rather than impartially, to facilitate access to research." In the committee's view, to fully advance representativeness in research, institutions and investigators must recognize the larger systemic context of their work, including historic abuses (e.g., the Tuskegee Syphilis Study) and the ongoing harms that shape the lived experiences of individuals, families, and communities. This new understanding rooted in social justice can position these stakeholders to better design participation pathways with people and communities at the center. Without a paradigm shift that looks beyond tactics and process-oriented changes, disparities in research access and inclusion will persist at the expense of minority population groups and the nation's public health (Gilmore-Bykovskyi et al., 2021).

References

AAMC (Association of American Medical Colleges). 2020. *The State of Women in Academic Medicine 2018–2019: Exploring Pathways to Equity.* Washington, DC: AAMC.

ACC (American College of Cardiology). 2022. Clinical Trials Research: Upping Your Game. Professional Development for Cardiologists and PhD Researchers Seeking to Succeed in Clinical Research. https://www.acc.org/CTR2020.

ACD (Advisory Committee to the NIH Director). 2015. "Participant Engagement and Health Equity Workshop Summary." ACD Precision Medicine Initiative Working Group Public Workshop, July 1–2, 2015, National Institutes of Health. http://www.nih.gov/precisionmedicine/2015-07-01-workshop-summary.pdf.

Adeyemi, O. F., A. T. Evans, and M. Bahk. 2009. HIV-infected adults from minority ethnic groups are willing to participate in research if asked. *AIDS Patient Care and STDs* 23(10):859–865.

Agnew, B. 1998. NIH embraces citizens' council to cool debate on priorities. *Science* 282(5386):18–19.

AHRQ (Agency for Healthcare Research and Quality). 2021. AHRQ Policy on the Inclusion of Priority Populations in Research. https://www.ahrq.gov/topics/individuals-special-healthcare-needs.html#:~:text=The%20AHRQ%20Policy%20on%20the%20Inclusion%20of%20Priority,and%20justification%20is%20provided%20that%20inclusion%20is%20inappropriate.

AIM (Alzheimer's Impact Movement). 2020. *Race, Ethnicity, and Alzheimer's: Fact Sheet.* https://www.alz.org/aaic/downloads2020/2020_Race_and_Ethnicity_Fact_Sheet.pdf.

Aitken, M., M. Kleinrick, E. Munoz, and U. Porwal. 2019. Orphan Drugs in the United States: Rare Disease Innovation and Cost Trends Through 2019. IQVIA. https://rarediseases.org/wp-content/uploads/2021/03/orphan-drugs-in-theunited-states-NRD-2020.pdf. Published December 3, 2020.

Alcaraz, K. I., N. L. Weaver, E. M. Andresen, K. Christopher, and M. W. Kreuter. 2011. The neighborhood voice: Evaluating a mobile research vehicle for recruiting African Americans to participate in cancer control studies. *Evaluation and the Health Professions* 34(3):336–348.

Alhajji, M., S. B. Bass, A. Nicholson, A. Washington, L. Maurer, D. M. Geynisman, and L. Fleisher. 2020. Comparing perceptions and decisional conflict towards participation in cancer clinical trials among African American patients who have and have not participated. *Journal of Cancer Education* 37(2):395–404.

Alsan, M., and S. Eichmeyer. 2021. Experimental evidence on the effectiveness of non-experts for improving vaccine demand. National Bureau of Economic Research Working Paper No. 28593. https://www.nber.org/papers/w28593.

Alsan, M., O. Garrick, and G. Graziani. 2019. Does diversity matter for health? Experimental evidence from Oakland. *American Economic Review* 109(12):4071–4111.

Ambrosius, W. T., K. M. Sink, C. G. Foy, D. R. Berlowitz, A. K. Cheung, W. C. Cushman, L. J. Fine, D. C. Goff, K. C. Johnson, A. A. Killeen, C. E. Lewis, S. Oparil, D. M. Reboussin, M. V. Rocco, J. K. Snyder, J. D. Williamson, J. T. Wright, and P. K. Whelton. 2014. The design and rationale of a multicenter clinical trial comparing two strategies for control of systolic blood pressure: The systolic blood pressure intervention trial (SPRINT). *Clinical Trials* 11(5):532–546.

Anderson, G. D. 2005. Sex and racial differences in pharmacological response: Where is the evidence? Pharmacogenetics, pharmacokinetics, and pharmacodynamics. *Journal of Women's Health* 14(1):19–29.

Arega, A., N. J. O. Birkmeyer, J. D. N. Lurie, T. Tosteson, J. Gibson, B. A. Taylor, T. S. Morgan, and J. N. Weinstein. 2006. Racial variation in treatment preferences and willingness to randomize in the spine patient outcomes research trial (SPORT). *Spine* 31(19):2263–2269.

Arnegard, M. E., L. A. Whitten, C. Hunter, and J. A. Clayton. 2020. Sex as a biological variable: A 5-year progress report and call to action. *Journal of Women's Health* 29(6):858–864.

Artiga, S., J. Kates, J. Michaud, L. Hill. 2021. Racial diversity within covid-19 vaccine clinical trials: Key questions and answers. *KFF Health Policy*. https://www.kff.org/racial-equity-and-health-policy/issue-brief/racial-diversity-within-covid-19-vaccine-clinical-trials-key-questions-and-answers.

Ashford, M. T., J. Eichenbaum, T. Williams, M. R. Camacho, J. Fockler, A. Ulbricht, D. Flenniken, D. Truran, R. S. Mackin, M. W. Weiner, and R. L. Nosheny. 2020. Effects of sex, race, ethnicity, and education on online aging research participation. *Alzheimers Dementia (N Y)*. 6(1):e12028.

Assmann, S. F., S. J. Pocock, L. E. Enos, and L. E. Kasten. 2000. Subgroup analysis and other (mis) uses of baseline data in clinical trials. *The Lancet* 355(9209):1064–1069.

AuYoung, M., P. Rodriguez Espinosa, W. T. Chen, P. Juturu, M. T. Young, A. Casillas, P. Adkins-Jackson, S. Hopfer, E. Kissam, A. K. Alo, R. A. Vargas, and A. F. Brown. 2022. Addressing racial/ethnic inequities in vaccine hesitancy and uptake: lessons learned from the California alliance against COVID-19. *Journal of Behavioral Medicine* 22:1–14.

Azar, K. M., M. R. Moreno, E. C. Wong, J. J. Shin, C. Soto, and L. P. Palaniappan. 2012. Accuracy of data entry of patient race/ethnicity/ancestry and preferred spoken language in an ambulatory care setting. *Health Services Research* 47:228–240.

Azfar-e-Alam, S., A. Sikorskii, C. W. Given, and B. Given. 2008. Early participant attrition from clinical trials: Role of trial design and logistics. *Clinical Trials* 5(4):328–335.

Baca, E., M. Garcia-Garcia, and A. Porras-Chavarino. 2004. Gender differences in treatment response to sertraline versus imipramine in patients with nonmelancholic depressive disorders. *Progress in Neuro-Psychopharmacol and Biological Psychiatry* 28(1):57–65.

Bahrampour, T., and T. Mellnik. 2021. Census data shows widening diversity; number of white people falls for first time. https://www.washingtonpost.com/dc-md-va/2021/08/12/census-data-race-ethnicity-neighborhoods/.

Bano, S., S. Akhter, and M. I. Afridi. 2004. Gender based response to fluoxetine hydrochloride medication in endogenous depression. *Journal of College Physicians and Surgeons Pakistan* 14(3):161–165.

Barnes, L. L., and D. A. Bennett. 2014. Alzheimer's disease in African Americans: Risk factors and challenges for the future. *Health Affairs* 33(4):580–586.

Bazan, I. S., and K. M. Akgün. 2021. COVID-19 healthcare inequity: Lessons learned from annual influenza vaccination rates to mitigate COVID-19 vaccine disparities. *Yale Journal of Biology and Medicine* 94(3):509–515.

Becker, G. S., T. J. Philipson, and R. R. Soares. 2005. The quantity and quality of life and the evolution of world inequality. *American Economic Review* 95(1):277–291.

REFERENCES

Becker, J. 2021. Excluding pregnant people from clinical trials reduces patient safety and autonomy. *Harvard Law* "Bill of Health" (blog), posted April 16, 2021. https://blog.petrieflom.law.harvard.edu/2021/04/16/pregnant-clinical-trials-safety-autonomy/.

Beglinger, C. 2008. Ethics related to drug therapy in the elderly. *Digestive Diseases* 26(1):28–31.

Behringer-Massera, S., T. Browne, G. George, S. Duran, A. Cherrington, M. D. McKee, and GRADE Research Group. 2019. Facilitators and barriers to successful recruitment into a large comparative effectiveness trial: A qualitative study. *Journal of Comparative Effectiveness Research* 8(10):815–826.

BeLue, R., K. D. Taylor-Richardson, J. Lin, A. T. Rivera, and D. Grandison. 2006. African Americans and participation in clinical trials: Differences in beliefs and attitudes by gender. *Contemporary Clinical Trials* 27(6):498–505.

Berkowitz, S. T., S. L. Groth, S. Gangaputra, and S. Patel. 2021. Racial/ethnic disparities in ophthalmology clinical trials resulting in US Food and Drug Administration drug approvals from 2000 to 2020. *JAMA Ophthalmology* 139(6):629–637.

Bieniasz, M. E., D. Underwood, J. Bailey, and M. T. Ruffin IV. 2003. Women's feedback on a chemopreventive trial for cervical dysplasia. *Applied Nursing Research* 16(1):22–28.

Bierer, B. E., S. A. White, L. Meloney, H. Ahmed, D. H. Strauss, and L. T. Clark. 2020. *Achieving Diversity, Inclusion, and Equity in Clinical Research: Guidance Document and Supplementary Toolkit*. Cambridge, MA: MRCT Center.

Bishop, W. P., J. A. Tiro, S. J. C. Lee, C. M. Bruce, and C. S. Skinner. 2011. Community events as viable sites for recruiting minority volunteers who agree to be contacted for future research. *Contemporary Clinical Trials* 32(3):369–371.

Bistricky, S. L., R. S. MacKin, J. P. Chu, and P. A. Areán. 2010. Recruitment of African Americans and Asian Americans with late-life depression and mild cognitive impairment. *American Journal of Geriatric Psychiatry* 18(8):734–742.

Blehar, M. C., C. Spong, C. Grady, S. F. Goldkind, L. Sahin, and J. A. Clayton. 2013. Enrolling pregnant women: Issues in clinical research. *Womens Health Issues* 23(1):e39–45.

Boehmer U., N. R. Kressin, D. R. Berlowitz, C. L. Christiansen, L. E. Kazis, and J. A. Jones. 2002. Self-reported vs administrative race/ethnicity data and study results. *American Journal of Public Health* 92(9):1471–1472.

Boise, L., L. Hinton, H. J. Rosen, and M. Ruhl. 2017. Will my soul go to heaven if they take my brain? Beliefs and worries about brain donation among four ethnic groups. *Gerontologist* 57(4):719–734.

Boissel, J.-P., F. Gueyffier, and M. Haugh. 1995. Response to "inclusion of women and minorities in clinical trials and the NIH revitalization act of 1993—the perspective of NIH clinical trialists." *Controlled Clinical Trials* 16(5):286–288.

Bonevski, B., M. Randell, C. Paul, K. Chapman, L. Twyman, J. Bryant, I. Brozek, and C. Hughes. 2014. Reaching the hard-to-reach: A systematic review of strategies for improving health and medical research with socially disadvantaged groups. *BMC Medical Research Methodology* 14:42.

Borno, H., D. J. George, L. E. Schnipper, F. Cavalli, T. Cerny, and S. Gillessen. 2019. All men are created equal: Addressing disparities in prostate cancer care. *American Society of Clinical Oncology Educational Book* 39:302–308.

Bothwell, L. E., J. A. Greene, S. H. Podolsky, and D. S. Jones. 2016. Assessing the gold standard—lessons from the history of RCTs. *New England Journal of Medicine* 374(22):2175–2181.

Boyatzis, R. 1998. *Transforming Qualitative Information: Thematic Analysis and Code Development*. Thousand Oaks, CA: SAGE Publications.

Brandt, A. M. 1978. Racism and research: The case of the Tuskegee syphilis study. *The Hastings Center Report* 8(6):21–29.

Brantnell, A., E. Baraldi, T. van Achterberg, and U. Winblad. 2015. Research funders' roles and perceived responsibilities in relation to the implementation of clinical research results: A multiple case study of Swedish research funders. *Implementation Science* 10(1):100.

Braunstein, J. B., N. S. Sherber, S. P. Schulman, E. L. Ding, and N. R. Powe. 2008. Race, medical researcher distrust, perceived harm, and willingness to participate in cardiovascular prevention trials. *Medicine* 87(1):1–9.

Brick-Hezeau, E. 2019. Who gets a say? Women and the AIDS crisis. *Women Leading Change: Case Studies on Women, Gender, and Feminism* 4(2). Tulane University Press.

Brown, D. R., M. N. Fouad, K. Basen-Engquist, and G. Tortolero-Luna. 2000. Recruitment and retention of minority women in cancer screening, prevention, and treatment trials. *Annals of Epidemiology* 10(8):S13–S21.

Brown, R. F., D. L. Cadet, R. H. Houlihan, M. D. Thomson, E. C. Pratt, A. Sullivan, and L. A. Siminoff. 2013. Perceptions of participation in a phase I, II, or III clinical trial among African American patients with cancer: What do refusers say? *Journal of Oncology Practice* 9(6):287–293.

Buchbinder, S. P., B. Metch, S. E. Holte, S. Scheer, A. Coletti, and E. Vittinghoff. 2004. Determinants of enrollment in a preventive HIV vaccine trial: Hypothetical versus actual willingness and barriers to participation. *Journal of Acquired Immune Deficiency Syndromes* 36(1):604–612.

Buck, D. S., D. Rochon, H. Davidson, and S. McCurdy. 2004. Involving homeless persons in the leadership of a health care organization. *Qualitative Health Research* 14(4):513–525.

Bussey-Jones, J., J. Garrett, G. Henderson, M. Moloney, C. Blumenthal, and G. Corbie-Smith. 2010. The role of race and trust in tissue/blood donation for genetic research. *Genetics in Medicine* 12(2):116–121.

Butler, J., S. C. Quinn, C. S. Fryer, M. A. Garza, K. H. Kim, and S. B. Thomas. 2013. Characterizing researchers by strategies used for retaining minority participants: Results of a national survey. *Contemporary Clinical Trials* 36(1):61–67.

Byrd, G. S., C. L. Edwards, V. A. Kelkar, R. G. Phillips, J. R. Byrd, D. S. Pim-Pong, T. D. Starks, A. L. Taylor, R. E. McKinley, Y.-J. Li, and M. Pericak-Vance. 2011. Recruiting intergenerational African American males for biomedical research studies: A major research challenge. *Journal of the National Medical Association* 103(6):480–487.

Byrne, J. J., A. M. Saucedo, and C. Y. Spong. 2020. Task force on research specific to pregnant and lactating women. *Seminars in Perinatology* 44(3):151226.

Byrne, M. M., S. L. Tannenbaum, S. Glück, J. Hurley, and M. Antoni. 2014. Participation in cancer clinical trials: Why are patients not participating? *Medical Decision Making* 34(1):116–126.

Calderon, J. L., R. S. Baker, H. Fabrega, J. G. Conde, R. D. Hays, E. Fleming, and K. Norris. 2006. An ethno-medical perspective on research participation: A qualitative pilot study. *MedGenMed* 8(2):23.

Califf, R. M., D. A. Zarin, J. M. Kramer, R. E. Sherman, L. H. Aberle, and A. Tasneem. 2012. Characteristics of clinical trials registered in ClinicalTrials.gov, 2007–2010. *JAMA* 307(17):1838–1847.

Canavan, C., S. Grossman, R. Kush, and J. Walker. 2006. Integrating recruitment into eHealth patient records. *Applied Clinical Trials* 0(0). https://www.appliedclinicaltrialsonline.com/view/integrating-recruitment-ehealth-patient-records.

Carapinha, R., C. M. McCracken, E. T. Warner, E. V. Hill, and J. Y. Reede. 2017. Organizational context and female faculty's perception of the climate for women in academic medicine. *Journal of Women's Health* 26(5):549–559.

Carcel, C., and M. Reeves. 2021. Under-enrollment of women in stroke clinical trials. *Stroke 2021* 52:452–457.

Carratala, S., and C. Maxwell. 2020. Health disparities by race and ethnicity. https://cdn.americanprogress.org/content/uploads/2020/05/06130714/HealthRace-factsheet.pdf?_ga=2.194956205.1806611089.1627327183-1860413079.1627327183.

Castillo-Mancilla, J. R., S. E. Cohn, S. Krishnan, M. Cespedes, M. Floris-Moore, G. Schulte, G. Pavlov, D. Mildvan, and K. Y. Smith; the ACTG Underrepresented Populations Survey Group. 2014. Minorities remain underrepresented in HIV/AIDS research despite access to clinical trials. *HIV Clinical Trials* 15(1):14–26.

CDC (Centers for Disease Control and Prevention). 1995. Policy on the inclusion of women and racial and ethnic minorities in externally awarded research. *Federal Register* 60:47947–47951.

REFERENCES

CDC. 1996. Inclusion of women and racial and ethnic minorities in research. C.F.R. Part 46.
CDC. 1997. CDC procedures for protection of human research participants. https://www.cdc.gov/os/integrity/hrpo/index.htm.
CDC. 2010. Inclusion of women and racial and ethnic minorities in research. CDC-GA-1996-01, C.F.R. Part 46.
CDC. 2011. Inclusion of persons under the age of 21 in research. CDC-GA-2006-01. https://www.cdc.gov/Maso/Policy/policy496.pdf.
CDC. 2020. Disparities. https://www.healthypeople.gov/2020/about/foundation-health-measures/Disparities.
Ceballos, R. M., S. Knerr, M. A. Scott, S. D. Hohl, R. C. Malen, H. Vilchis, and B. Thompson. 2014. Latino beliefs about biomedical research participation: A qualitative study on the U.S.-Mexico border. *Journal of Empirical Research on Human Research Ethics* 9(4):10–21.
Census (U.S. Census Bureau). 2018. Hispanic population to reach 111 million by 2060. https://www.census.gov/library/visualizations/2018/comm/hispanic-projected-pop.html.
Census. 2019. The American Community Survey 2005–2019. https://www.census.gov/programs-surveys/acs/.
Chen, B., H. Jin, Z. Yang, Y. Qu, H. Weng, and T. Hao. 2019. An approach for transgender population information extraction and summarization from clinical trial text. *BMC Medical Informatics and Decision Making* 19(Suppl. 2):62.
Chen, M. S. Jr., P. N. Lara, J. H. Dang, D. A. Paterniti, and K. Kelly. 2014. Twenty years post-NIH Revitalization Act: Enhancing minority participation in clinical trials (EMPaCT): Laying the groundwork for improving minority clinical trial accrual: Renewing the case for enhancing minority participation in cancer clinical trials. *Cancer* 120:1091–1096.
Choi, E., G. J. Heo, Y. Song, and H.-R. Han. 2016. Community health worker perspectives on recruitment and retention of recent immigrant women in a randomized clinical trial. *Family & Community Health* 39(1):53–61.
Clark, H. 2013. Towards a more robust investment framework for health. *Lancet* 382(9908):e36–37.
Clark, L. T., L. Watkins, I. L. Piña, M. Elmer, O. Akinboboye, M. Gorham, B. Jamerson, C. McCullough, C. Pierre, A. B. Polis, G. Puckrein, and J. M. Regnante. 2019. Increasing diversity in clinical trials: Overcoming critical barriers. *Current Problems in Cardiology* 44(5):148–172.
Clayton, J. A. 2021. *The NIH SABV policy: E-learning opportunities and a symposium provide guidance and inspiration*. https://nexus.od.nih.gov/all/2021/04/12/the-nih-sabv-policy-e-learning-opportunities-and-a-symposium-provide-guidance-and-inspiration/.
Clayton, J. A., and F. S. Collins. 2014. Policy: NIH to balance sex in cell and animal studies. *Nature News* 509(7500):282.
CMS (Centers for Medicare & Medicaid Services). n.d. Coverage with Evidence Development. https://www.cms.gov/Medicare/Coverage/Coverage-with-Evidence-Development.
CMS. 2014b. Medicare Coverage Database: Transcatheter Mitral Valve Repair (TMVR). CAG-00438N. https://www.cms.gov/medicare-coverage-database/view/ncacal-decision-memo.aspx?proposed=N&NCAId=273.
CMS. 2000. 2000 National Coverage Determination (NCD). Routine Costs in Clinical Trials. https://www.cms.gov/medicare-coverage-database/view/ncd.aspx?ncdid=1&ncdver=1&fromdb=true.
CMS. 2007. 2007 National Coverage Determination (NCD). Routine Costs in Clinical Trials. https://www.cms.gov/medicare-coverage-database/view/ncd.aspx?NCDId=1&fromdb=true.
CMS. 2014a. Guidance for the public, industry, and CMS staff: Coverage with evidence development. https://www.cms.gov/medicare-coverage-database/view/medicare-coverage-document.aspx?MCDId=27.
Coakley, M., E. O. Fadiran, L. J. Parrish, R. A. Griffith, E. Weiss, and C. Carter. 2012. Dialogues on diversifying clinical trials: Successful strategies for engaging women and minorities in clinical trials. *Journal of Women's Health* 21(7):713–716.

Coburn, A., A. C. Mackinney, T. McBride, K. Mueller, R. Slifkin, and M. Wakefield. 2017. *Choosing Rural Definitions: Implications for Health Policy.* Issue Brief No. 2. Omaha, NE: Rural Policy Research Institute Health Panel. https://digitalcommons.usm.maine.edu/insurance/60/.

Cohen, J. C., E. Boerwinkle, T. H. Mosley, and H. H. Hobbs. 2006. Sequence variations in PCSK9, low LDL, and protection against coronary heart disease. *New England Journal of Medicine* 354(12):1264–1272.

Collins, F. S., and H. Varmus. 2015. A new initiative on precision medicine. *New England Journal of Medicine* 372(9):793–795.

COMMITTEE ON DRUGS, Kathleen A. Neville, Daniel A.C. Frattarelli, Jeffrey L. Galinkin, Thomas P. Green, Timothy D. Johnson, MMM, Ian M. Paul, John N. Van Den Anker. 2014. Off-Label Use of Drugs in Children. *Pediatrics* 133 (3): 563–567.

Corbie-Smith, G., S. B. Thomas, and D. M. St. George. 2002. Distrust, race, and research. *Archives of Internal Medicine* 162(21):2458–2463.

Cornwall, A., and R. Jewkes. 1995. What is participatory research? *Social Science & Medicine* 41(12):1667–1676.

Coronado, G. D., S. Ondelacy, Y. Schwarz, C. Duggan, J. W. Lampe, and M. L. Neuhouser. 2012. Recruiting underrepresented groups into the Carbohydrates and Related Biomarkers (CARB) cancer prevention feeding study. *Contemporary Clinical Trials* 33(4):641–646.

Cotton, P. 1990. Examples abound of gaps in medical knowledge because of groups excluded from scientific study. *JAMA* 263(8):1051, 1055.

Crawley, F. P., R. Kurz, and H. Nakamura. 2003. Testing medications in children. *New England Journal of Medicine* 348(8):763–764; author's reply 763–764.

Crenshaw, K. 2017. *On Intersectionality: Essential Writings.* New York: The New Press.

Crocker, J. C., I. Ricci-Cabello, A. Parker, J. A. Hirst, A. Chant, S. Petit-Zeman, D. Evans, and S. Rees. 2018. Impact of patient and public involvement on enrolment and retention in clinical trials: Systematic review and meta-analysis. *BMJ*:k4738.

Cummins, N. W., J. Neuhaus, H. Chu, J. Neaton, C. Wyen, J. K. Rockstroh, D. J. Skiest, M. A. Boyd, S. Khoo, M. Rotger, A. Telenti, R. Weinshillboum, and A. D. Badley. 2015. Investigation of efavirenz discontinuation in multi-ethnic populations of HIV-positive individuals by genetic analysis. *eBioMedicine* 2(7):706–712.

Curry, L., and J. Jackson. 2003. The science of including older ethnic and racial group participants in health-related research. *Gerontologist* 43(1):15–17.

DHEW (U.S. Department of Health, Education, and Welfare). 1979. National Commission for the Protection of Human Subjects of Biomedical and Behavioral Research. *The Belmont Report: Ethical Principles and Guidelines for the Protection of Human Subjects of Research.* Washington, DC: Office for Human Research Protections, Department of Health and Human Services.

Dilworth-Anderson, P. 2011. Introduction to the science of recruitment and retention among ethnically diverse populations. *Gerontologist* 51(Suppl. 1):S1–S4.

DiPietro, N. A., and K. A. Liu. 2016. Women's involvement in clinical trials: Historical perspective and future implications. *Pharmacy Practice (Granada)* 14(1):5.

Dohan, D., and D. Schrag. 2005. Using navigators to improve care of underserved patients. *Cancer* 104(4):848–855.

DOL (Department of Labor). 2020. Earnings. https://www.dol.gov/agencies/wb/data/earnings.

Domecq, J. P., G. Prutsky, T. Elraiyah, Z. Wang, M. Nabhan, N. Shippee, J. P. Brito, K. Boehmer, R. Hasan, B. Firwana, P. Erwin, D. Eton, J. Sloan, V. Montori, N. Asi, A. M. Abu Dabrh, and M. H. Murad. 2014. Patient engagement in research: A systematic review. *BMC Health Services Research* 14(1):89.

Drozda, K., S. Wong, S. R. Patel, A. P. Bress, E. A. Nutescu, R. A. Kittles, and L. H. Cavallari. 2015. Poor warfarin dose prediction with pharmacogenetic algorithms that exclude genotypes important for African Americans. *Pharmacogenetics and Genomics* 25(2):73–81.

REFERENCES

Duma, N., J. V. Aguilera, J. Paludo, C. Haddox, M. G. Velez, Y. Wang, K. Leventakos, J. Hubbard, A. Mansfield, R. Go, and A. Adjei. 2018. Representation of minorities and women in oncology clinical trials: Review of the past 14 years. *Journal of Oncology Practice* 14(1).

Durant, R. W., J. A. Wenzel, I. C. Scarinci, D. A. Paterniti, M. N. Fouad, T. C. Hurd, and M. Y. Martin. 2014. Perspectives on barriers and facilitators to minority recruitment for clinical trials among cancer center leaders, investigators, research staff, and referring clinicians: enhancing minority participation in clinical trials (EMPaCT). *Cancer* 120 (0 7): 1097–1105. https://doi.org/10.1002/cncr.28574.

Eggly, S., E. Barton, A. Winckles, L. A. Penner, and T. L. Albrecht. 2015. A disparity of words: Racial differences in oncologist-patient communication about clinical trials. *Health Expectations* 18(5):1316–1326.

Ejiogu, N., J. H. Norbeck, M. A. Mason, B. C. Cromwell, A. B. Zonderman, and M. K. Evans. 2011. Recruitment and retention strategies for minority or poor clinical research participants: Lessons from the healthy aging in neighborhoods of diversity across the life span study. *Gerontologist* 51(Suppl. 1):S33–S45.

Emanuel, E. J. 2005. Undue inducement: Nonsense on stilts? *American Journal of Bioethics* 5(5):9–13; discussion W18-11, W17.

Ermini Leaf, D., B. Tysinger, D. P. Goldman, and D. N. Lakdawalla. 2021. Predicting quantity and quality of life with the future elderly model. *Health Economics* 30(Suppl. 1):52–79.

Eshera, N., H. Itana, L. Zhang, G. Soon, and E. O. Fadiran. 2015. Demographics of clinical trials participants in pivotal clinical trials for new molecular entity drugs and biologics approved by FDA from 2010 to 2012. *American Journal of Therapeutics* 22(6):435–455.

Evans, M. K., J. M. Lepkowski, N. R. Powe, T. LaVeist, M. F. Kuczmarski, and A. B. Zonderman. 2010. Healthy aging in neighborhoods of diversity across the life span (HANDLS): Overcoming barriers to implementing a longitudinal, epidemiologic, urban study of health, race, and socioeconomic status. *Ethnicity and Disease* 20(3):267–275.

Executive Office of the President. 2021. Executive Order No. 13985: Advancing racial equity and support for underserved communities through the federal government. Washington, DC: The White House.

FDA (Food and Drug Administration). n.d. Descovy prescribing information. https://www.accessdata.fda.gov/drugsatfda_docs/label/2021/208215s017lbl.pdf.

FDA. 1977. General Considerations for the Clinical Evaluation of Drugs. https://www.fda.gov/regulatory-information/search-fda-guidance-documents/general-considerations-clinical-evaluation-drugs.

FDA. 1979. Labeling and Prescription Drug Advertising: Content and Format for Labeling for Human Prescription Drugs. *Federal Register* 44(124):434–467.

FDA. 1985. Content and Format of a New Drug Application. 21 C.F.R. 314.50 (d)(5)(v).

FDA. 1987. Guideline for the Format and Content of the Nonclinical Pharmacology/Toxicology Section of an Application.

FDA. 1988. Guideline for the Format and Content of the Clinical and Statistical Sections of New Drug Applications. https://www.fda.gov/regulatory-information/search-fda-guidance-documents/format-and-content-nonclinical-pharmacologytoxicology-section-application.

FDA. 1989. Guideline for the Study of Drugs Likely to be Used in the Elderly. https://www.fda.gov/regulatory-information/search-fda-guidance-documents/study-drugs-likely-be-used-elderly.

FDA. 1993a. Guideline for the Study and Evaluation of Gender Differences in the Clinical Evaluation of Drugs; Notice. *Federal Register* 58(139):39406–39416.

FDA. 1993b. Guidance for Industry. https://www.fda.gov/media/72504/download.

FDA. 1998. Guidance for industry on population pharmacokinetics. *Federal Register* 68(28):1854–1862.

FDA. 1999. Guidance for industry on population pharmacokinetics. *Federal Register* 64(27):6663–6664.

FDA. 2002. Guidance for industry on establishing pregnancy exposure registries. *Federal Register* 67:59528.

FDA. 2004. Guidance for industry on pharmacokinetics in pregnancy-Study design, data analysis, and impact on dosing and labeling. *Federal Register* 69:63402–63403.

FDA. 2011a. Dialogues on diversifying clinical trials. September 22–23, 2011, Washington, DC.

FDA. 2011b. Guidance for industry: reproductive and developmental toxicities—integrating study results to assess concerns. https://www.fda.gov/regulatory-information/search-fda-guidance-documents/reproductive-and-developmental-toxicities-integrating-study-results-assess-concerns.

FDA. 2012. FDA Safety and Innovation Act (FDASIA) Section 907: Inclusion of Demographic Subgroups in Clinical Trials. https://www.fda.gov/regulatory-information/food-and-drug-administration-safety-and-innovation-act-fdasia/fdasia-section-907-inclusion-demographic-subgroups-clinical-trials#:~:text=Sec.%20907%20of%20the%20Food%20and%20Drug%20Administration,devices%2C%20submitted%20to%20the%20agency%20for%20marketing%20approval%3A.

FDA. 2013. Collection, analysis and availability of demographic subgroup data for FDA-approved medical products. https://www.fda.gov/files/about%20fda/published/Collection--Analysis--and-Availability-of-Demographic-Subgroup-Data-for-FDA-Approved-Medical-Products.pdf. *or* Washington, DC: Department of Health and Human Services.

FDA. 2014a. Evaluation of sex-specific data in medical device clinical studies: Guidance for industry and FDA staff. Docket No. FDA-2011-D-0817. Washington, DC: Department of Health and Human Services.

FDA. 2014b. *FDA Action Plan to Enhance the Collection and Availability of Demographic Subgroup Data.* Washington, DC: Food and Drug Administration. https://www.fda.gov/media/89307/download.

FDA. 2016. Collection of race and ethnicity data in clinical trials: Guidance for industry and Food and Drug Administration staff. Docket No. FDA-2016-D-3561. Washington, DC: Food and Drug Administration. https://www.fda.gov/regulatory-information/search-fda-guidance-documents/collection-race-and-ethnicity-data-clinical-trials.

FDA. 2017. Evaluation and Reporting of Age-, Race-, and Ethnicity –Specific Data in Medical Device Clinical Studies. *Federal Register* 82:42819–42821.

FDA. 2018. Pregnant Women: Scientific and Ethical Considerations for Inclusion in Clinical Trials. *Federal Register* 83:15161–15162.

FDA. 2019. Meeting of the Antimicrobial Drugs Advisory Committee. https://www.fda.gov/advisory-committees/antimicrobial-drugs-advisory-committee-formerly-known-anti-infective-drugs-advisory-committee/2019-meeting-materials-antimicrobial-drugs-advisory-committee-formerly-known-anti-infective-drugs for list of dates.

FDA. 2020a. 2015–2019 Drug Trials Snapshots Summary Report: Five-year Summary and Analysis of Clinical Trial Participation and Demographics. https://www.fda.gov/media/143592/download. https://www.fda.gov/media/143592/download (accessed July 8th, 2021).

FDA. 2020b. Statement: FDA offers guidance to enhance diversity in clinical trials, encourage inclusivity in medical product development. Press Announcement, November 9, 2020. https://www.fda.gov/news-events/press-announcements/fda-offers-guidance-enhance-diversity-clinical-trials-encourage-inclusivity-medical-product.

FDA. 2020c. Enhancing the Diversity of Clinical Trial Populations – Eligibility Criteria, Enrollment Practices, and Trial Designs. Guidance for Industry. https://www.fda.gov/regulatory-information/search-fda-guidance-documents/enhancing-diversity-clinical-trial-populations-eligibility-criteria-enrollment-practices-and-trial.

FDA. 2020d. Are You Concerned about Clinical Trial Enrollment and Representation? SBIA Webinar, December 16, 2020. https://sbiaevents.com/files2/Diversity-Clinical-Trials-Slides.pdf.

FDA. 2021a. Guidances. https://www.fda.gov/industry/fda-basics-industry/guidances.

FDA. 2021b. Drug Trials Snapshots. https://www.fda.gov/drugs/drug-approvals-and-databases/drug-trials-snapshots.

Findley, S., S. Matos, A. Hicks, J. Chang, and D. Reich. 2014. Community health worker integration into the health care team accomplishes the triple aim in a patient-centered medical home: A Bronx tale. *Journal of Ambulatory Care Management* 37(1):82–91.

Fisher, C. B., and B. Mustanski. 2014. Reducing health disparities and enhancing the responsible conduct of research involving LGBT youth. *Hastings Center Report* 44(Suppl. 4):S28–S31. doi:10.1002/hast.367. PMID: 25231783; PMCID: PMC4617525.

Fisher, C. B., L. I. Puri, K. Macapagal, L. Feuerstahler, J. R. Ahn, and B. Mustanski. 2021. Competence to consent to oral and injectable PrEP trials among adolescent males who have sex with males. *AIDS and Behavior* 25:1606–1618. https://doi.org/10.1007/s10461-020-03077-9.

Fisher, K. 2019. Academic health centers save millions of lives. American Association of Medical Colleges. https://www.aamc.org/news-insights/academic-health-centers-save-millions-lives.

Fleurence, R. L., L. H. Curtis, R. M. Califf, R. Platt, J. V. Selby, and J. S. Brown. 2014. Launching PCORnet, a national patient-centered clinical research network. *Journal of the American Medical Informatics Association* 21(4):578–582.

Flores, L. E., W. R. Frontera, M. P. Andrasik, C. del Rio, A. Mondríguez-González, S. A. Price, E. M. Krantz, S. A. Pergam, and J. K. Silver. 2021. Assessment of the inclusion of racial/ethnic minority, female, and older individuals in vaccine clinical trials. *JAMA Network Open* 4(2):e2037640–e2037640.

Fontanarosa, P. B., A. Flanagin, J. Z. Ayanian, R. O. Bonow, N. M. Bressler, D. Christakis, M. L. Disis, S. A. Josephson, M. R. Kibbe, D. Öngür, J. F. Piccirillo, R. F. Redberg, F. P. Rivara, K. Shinkai, and C. W. Yancy. 2021. Equity and the JAMA Network. *JAMA* 326(7):618–620.

Ford, J. G., M. W. Howerton, G. Y. Lai, T. L. Gary, S. B. Mid, M. C. Gibbons, J. Tilburt, C. Baffi, T. P. Tanpitukpongse, R. F. Wilson, N. R. Powe, and E. B. Bass. 2008. Barriers to recruiting underrepresented populations to cancer clinical trials: A systematic review. *Cancer* 112(2):228–242.

Ford, M., L. Siminoff, E. Pickelsimer, A. Mainous, D. Smith, V. Diaz, L. Soderstrom, M. Jefferson, and B. Tilley. 2013. Unequal burden of disease, unequal participation in clinical trials: Solutions from African American and Latino community members. *Health & Social Work* 38(1):29–38.

Forsythe, L. P., K. L. Carman, V. Szydlowski, L. Fayish, L. Davidson, D. H. Hickam, C. Hall, G. Bhat, D. Neu, L. Stewart, M. Jalowsky, N. Aronson, and C. U. Anyanwu. 2019. Patient engagement in research: Early findings from the Patient-Centered Outcomes Research Institute. *Health Affairs* 38(3):359–367.

Freedman, L. S., R. Simon, M. A. Foulkes, L. Friedman, N. L. Geller, D. J. Gordon, and R. Mowery. 1995. Perspective of NIH clinical trialists. *Controlled Clinical Trials* 16:277–285.

Freeman, H. P., B. J. Muth, and J. F. Kerner. 1995. Expanding access to cancer screening and clinical follow-up among the medically underserved. *Cancer Practice* 3(1):19–30.

Frew, P., D. Saint-Victor, M. Isaacs, S. Kim, G. Swamy, J. Sheffield, K. Edwards, T. Villafana, O. Kamagate, and K. Ault. 2014. Recruitment and retention of pregnant women into clinical research trials: An overview of challenges, facilitators, and best practices. *Clinical Infectious Diseases: An Official Publication of the Infectious Diseases Society of America* 59(Suppl. 7).

Frey, W. 2019. Six Maps that Reveal America's Expanding Racial Diversity. https://www.brookings.edu/research/americas-racial-diversity-in-six-maps/.

Friedman, D. J., B. B. Cohen, A. R. Averbach, and J. M. Norton. 2000. Race/ethnicity and OMB directive 15: Implications for state public health practice. *American Journal of Public Health* 90(11):1714.

Friedman, S. H., C. O. Cunningham, J. Lin, L. B. Haramati, and J. M. Levsky. 2020. Having a primary care provider is the strongest predictor of successful follow-up of participants in a clinical trial. *Journal of the American Board of Family Medicine* 33(3):431–439.

Frosch, D. L., and M. Tai-Seale. 2014. R-e-s-p-e-c-t—what it means to patients. *Journal of General Internal Medicine* 29(3):427–428.

Gadegbeku, C. A., P. K. Stillman, M. D. Huffman, J. S. Jackson, J. W. Kusek, and K. A. Jamerson. 2008. Factors associated with enrollment of African Americans into a clinical trial: Results from the African American study of kidney disease and hypertension. *Contemporary Clinical Trials* 29(6):837–842.

GAO (Government Accountability Office). 1992. FDA needs to ensure more study of gender differences in prescription drug testing. https://www.fda.gov/media/75639/download.

GAO. 2001. Women sufficiently represented in new drug testing but FDA oversight needs improvement. https://www.fda.gov/media/75566/download.

GAO. 2015. *National Institutes of Health: Better Oversight Needed to Help Ensure Continued Progress Including Women in Health Research*. Report to Congressional Requesters (GAO-16-13). Washington, DC: U.S. Government Accountability Office.

GAO. 2018a. *Indian Health Service: Agency Faces Ongoing Challenges Filling Provider Vacancies*. Report to Congressional Requesters (GAO-18-580). https://www.gao.gov/products/gao-18-580.

GAO. 2018b. Orphan Drugs: FDA Could Improve Designation Review Consistency; Rare Disease Drug Development Challenges Continue. Report to Congress. https://www.gao.gov/assets/gao-19-83.pdf.

Garber, M., B. H. Hanusa, G. E. Switzer, J. Mellors, and R. M. Arnold. 2007. HIV-infected African Americans are willing to participate in HIV treatment trials. *Journal of General Internal Medicine* 22(1):17–42.

Garcia, M., S. L. Mulvagh, C. N. Merz, J. E. Buring, and J. E. Manson. 2016. Cardiovascular disease in women: Clinical perspectives. *Circulation Research* 118(8):1273–1293.

Garrison, N. A. 2013. Genomic justice for Native Americans. *Science, Technology, & Human Values* 38(2):201–223.

Geller, S. E., A. Koch, B. Pellettieri, and M. Carnes. 2011. Inclusion, analysis, and reporting of sex and race/ethnicity in clinical trials: Have we made progress? *Journal of Women's Health* 20(3):315–320.

Geller, S. E., A. R. Koch, P. Roesch, A. Filut, E. Hallgren, and M. Carnes. 2018. The more things change, the more they stay the same: A study to evaluate compliance with inclusion and assessment of women and minorities in randomized controlled trials. *Academic Medicine: Journal of the Association of American Medical Colleges* 93(4):630.

Genetech Inc. 2020. A Long-Term Safety Extension of Studies ABE4869g and ABE4955g in Participants with Mild to Moderate Alzheimer's Disease Treated with Crenezumab. https://www.clinicaltrials.gov/ct2/show/results/NCT01723826.

George, S., N. Duran, and K. Norris. 2014. A systematic review of barriers and facilitators to minority research participation among African Americans, Latinos, Asian Americans, and Pacific Islanders. *American Journal of Public Health* 104(2):e16–e31.

Giarelli, E., D. W. Bruner, E. Nguyen, S. Basham, P. Marathe, D. Dao, T. N. Huynh, J. Cappella, and G. Nguyen. 2011. Research participation among Asian American women at risk for cervical cancer: Exploratory pilot of barriers and enhancers. *Journal of Immigrant and Minority Health* 13(6):1055–1068.

Gilbert, A. L., Knopf, A. S., Fortenberry, J. D., Hosek, S. G., Kapogiannis, B. G., Zimet, G. D. 2015. Adolescent self-consent for biomedical human immunodeficiency virus prevention research. *Journal of Adolescent Health* 57(1): 113-119.

Gilmore-Bykovskyi, A., J. D. Jackson, and C. H. Wilkins. 2021. The urgency of justice in research: Beyond COVID-19. *Trends in Molecular Medicine* 27(2):97–100.

Gilmore-Bykovskyi, A., Y. Jin, C. Gleason, S. Flowers-Benton, L. Block, P. Dilworth-Anderson, L. Barnes, M. Shah, and M. Zuelsdorff. 2019. Recruitment and retention of underrepresented populations in Alzheimer's disease research: A systematic review. *Alzheimer's & Dementia (N Y)* 5:751–770.

Giuliano, A., M. Papenfuss, J. D. de Zapien, S. Tilousi, and L. Nuvayestewa. 1998. Breast cancer screening among southwest American Indian women living on-reservation. *Preventive Medicine* 27(1):135–143.

REFERENCES

Giuliano, A. R., N. Mokuau, C. Hughes, G. Tortolero-Luna, B. Risendal, R. C. S. Ho, T. E. Prewitt, and W. J. McCaskill-Stevens. 2000. Participation of minorities in cancer research: The influence of structural, cultural, and linguistic factors. *Annals of Epidemiology* 10(8):S22–S34.

Glantz, L. H. 1998. Research with children. *American Journal of Law & Medicine* 24(2–3):213–244.

Glasgow, R., A. Huebschmann, and R. Brownson. 2018. Expanding the consort figure: Increasing transparency in reporting on external validity. *American Journal of Preventive Medicine* 55(3):422–430.

GlobalData. 2021. Increased use of virtual trials has contributed to improved patient accrual rates, says GlobalData. https://www.globaldata.com/increased-use-virtual-trials-contributed-improved-patient-accrual-rates-says-globaldata/.

Gluck, M. A., A. Shaw, and D. Hill. 2018. Recruiting older African Americans to brain health and aging research through community engagement: Lessons from the African-American brain health initiative at Rutgers University-Newark. *Generations* 42(2):78–82.

Goldman, D., P.-C. Michaud, D. Lakdawalla, Y. Zheng, A. Gailey, and I. Vaynman. 2010. The fiscal consequences of trends in population health. *National Tax Journal* 63(2):307.

Goldman, D. P., D. Cutler, J. W. Rowe, P. C. Michaud, J. Sullivan, D. Peneva, and S. J. Olshansky. 2013. Substantial health and economic returns from delayed aging may warrant a new focus for medical research. *Health Affairs (Millwood)* 32(10):1698–1705.

Goldman, D. P., and P. R. Orszag. 2014. The growing gap in life expectancy: Using the future elderly model to estimate implications for Social Security and Medicare. *American Economic Review* 104(5):230–233.

Goldman, D. P., B. Shang, J. Bhattacharya, A. M. Garber, M. Hurd, G. F. Joyce, D. N. Lakdawalla, C. Panis, and P. G. Shekelle. 2005. Consequences of health trends and medical innovation for the future elderly: When demographic trends temper the optimism of biomedical advances, how will tomorrow's elderly fare? *Health Affairs* 24(Suppl. 2):W5-R5–W5-R17.

Goldman, D. P., Y. Zheng, F. Girosi, P.-C. Michaud, S. J. Olshansky, D. Cutler, and J. W. Rowe. 2009. The benefits of risk factor prevention in Americans aged 51 years and older. *American Journal of Public Health* 99(11):2096–2101.

Gollin, L. X., R. C. Harrigan, J. L. Calderon, J. Perez, and D. Easa. 2005. Improving Hawaiian and Filipino involvement in clinical research opportunities: Qualitative findings from Hawai'i. *Ethnicity & Disease* 15(4 Suppl. 5):S5-111–119.

Gong, I., N. Tan, S. Ali, G. Lebovic, M. Mamdani, S. Goodman, D. Ko, A. Laupacis, and A. Yan. 2019. Temporal trends of women enrollment in major cardiovascular randomized clinical trials. *Canadian Journal of Cardiology* 35(5).

Green-Harris, G., S. L. Coley, R. L. Koscik, N. C. Norris, S. L. Houston, M. A. Sager, S. C. Johnson, and D. F. Edwards. 2019. Addressing disparities in Alzheimer's disease and African-American participation in research: An asset-based community development approach. *Frontiers in Aging Neuroscience* 11:125.

Greer, T. 2015. Landmark study shows intensive blood pressure management may save lives. *UAB News*, Research & Innovation, September 11, 2015. https://www.uab.edu/news/research/item/6486-landmark-study-shows-intensive-blood-pressure-management-may-save-lives.

Griffith, D. M., E. M. Bergner, A. S. Fair, and C. H. Wilkins. 2021. Using mistrust, distrust, and low trust precisely in medical care and medical research advances health equity. *American Journal of Preventive Medicine* 60(3):442–445.

Group, S. R. 2015. A randomized trial of intensive versus standard blood-pressure control. *New England Journal of Medicine* 373(22):2103–2116.

Guadagnolo, B. A., D. G. Petereit, P. Helbig, D. Koop, P. Kussman, E. F. Dunn, and A. Patnaik. 2009. Involving American Indians and medically underserved rural populations in cancer clinical trials. *Clinical Trials* 6(6):610–617.

Guthrie, S., J. Krapels, C. A. Lichten, and S. Wooding. 2016. *100 Metrics to Assess and Communicate the Value of Biomedical Research: An Ideas Book*. Santa Monica, CA: RAND Corporation.

Haidich, A., and J. Ioannidis. 2001. Patterns of patient enrollment in randomized controlled trials. *Journal of Clinical Epidemiology* 54(9):877–883.

Hamad, R., T. T. Nguyen, J. Bhattacharya, M. M. Glymour, and D. H. Rehkopf. 2019. Educational attainment and cardiovascular disease in the United States: A quasi-experimental instrumental variables analysis. *PLoS Med.* 16(6):e1002834.

Hamad, R., J. Penko, D. S. Kazi, P. Coxson, D. Guzman, P. C. Wei, A. Mason, E. A. Wang, L. Goldman, K. Fiscella, and K. Bibbins-Domingo. 2020. Association of low socioeconomic status with premature coronary heart disease in US adults. *JAMA Cardiology* 5(8):899–908.

Hamel L.M., L. A. Penner, T. L. Albrecht, E. Heath, C. K. Gwede, and S. Eggly. 2016. Barriers to clinical trial enrollment in racial and ethnic minority patients with cancer. *Cancer Control* 23(4):327–337. doi:10.1177/107327481602300404. PMID: 27842322; PMCID: PMC5131730.

Hamilton, B. E., J. A. Martin, and M. J. K. Osterman. 2021. *Births: Provisional Data for 2020*. National Center for Health Statistics, Division of Vital Statistics. NVSS Vital Statistics Rapid Release, Report No. 012. https://stacks.cdc.gov/view/cdc/104993.

Hardie, G., R. Liu, J. Darden, and W. M. Gold. 2010. Ethnic differences in methacholine responsiveness and word descriptors in African Americans, Hispanic-Mexican Americans, Asian-Pacific Islanders, and Whites with mild asthma. *Journal of Asthma* 47(4):388–396.

Hardie, G. E., R. Liu, J. Darden, and W. M. Gold. 2011. Recruitment of asthmatic ethnic minorities into a methacholine research study: Factors influencing participation. *Journal of the National Medical Association* 103(2):138–149.

Harrington, R., M. Hanna, E. Oehrlein, R. Camp, R. Wheeler, C. Cooblall, T. Tesoro, A. Scott, R. Gizycki, F. Nguyen, A. Hareendran, D. Patrick, and E. Perfetto. 2020. Defining patient engagement in research: Results of a systematic review and analysis: Report of the ISPOR patient-centered special interest group. *Value in Health* 23(6):677–688.

Harris, D. J., and P. S. Douglas. 2000. Enrollment of women in cardiovascular clinical trials funded by the National Heart, Lung, and Blood Institute. *New England Journal of Medicine* 343(7):475–480.

Hattam, V. 2005. Ethnicity & the boundaries of race: Rereading Directive 15. *Daedalus* 134(1):61–69.

Haynes-Maslow, L., P. Godley, L. Dimartino, B. White, J. Odom, A. Richmond, and W. Carpenter. 2014. African American women's perceptions of cancer clinical trials. *Cancer Medicine* 3(5):1430–1439.

Helmuth, L. 2000. Reports see progress, problems, in trials. *Science* 288(5471):1562–1563.

Hersh, W. R., M. G. Weiner, P. J. Embi, J. R. Logan, P. S. Payne, E. V. Bernstam, H. P. Lehmann, G. Hripcsak, T. H. Hartzog, J. J. Cimino, and J. H. Saltz. 2013. Caveats for the use of operational electronic health record data in comparative effectiveness research. *Medical Care* 51(8 Suppl. 3):S30–S37.

HHS (Department of Health and Human Services). 1983. Additional protections for children involved as subjects in research—Department of Health and Human Services. Final rule. *Federal Register* 48(46):9814–9820.

HHS. 1985. Office on Women's Health. *Report of the Public Health Service Task Force on Women's Health Issues*. Washington, DC: Department of Health and Human Services.

HHS. 1994. Part R. Statement of organization, functions, and delegations of authority. 86 FR 48737. https://www.federalregister.gov/documents/2021/08/31/2021-18075/statement-of-organization-functions-and-delegations-of-authority.

HHS. 2010. Office of Inspector General. *Challenges to FDA's Ability to Monitor and Inspect Foreign Clinical Trials*. Washington, DC: Department of Health and Human Services.

HHS. 2017. Federal policy for the protection of human subjects. *Federal Register* 82:7149–7274.

HHS. 2020. *Civil Money Penalties Relating to the ClinicalTrials.gov Data Bank*. Washington, DC: Department of Health and Human Services.

HHS. 2022a. Office of Minority Health. Profile: American Indian/Alaska Native. https://minorityhealth.hhs.gov/omh/browse.aspx?lvl=3&lvlid=62.

HHS. 2022b. *Breastfeeding: Surgeon General's Call to Action Fact Sheet.* https://www.hhs.gov/surgeongeneral/reports-and-publications/breastfeeding/factsheet/index.html.
Hill, C., E. Pérez-Stable, N. Anderson, and M. Bernard. 2015. The National Institute on Aging health disparities research framework. *Ethnicity & Disease* 25(3).
Hillyer, G. C., M. Beauchemin, D. L. Hershman, M. Kelsen, F. L. Brogan, R. Sandoval, K. M. Schmitt, A. Reyes, M. B. Terry, A. B. Lassman, and G. K. Schwartz. 2020. Discordant attitudes and beliefs about cancer clinical trial participation between physicians, research staff, and cancer patients. *Clinical Trials* 17(2):184–194.
Hooper, M. W., T. Asfar, M. Unrod, A. Dorsey, J. B. Correa, K. O. Brandon, V. N. Simmons, M. A. Antoni, T. Koru-Sengul, D. J. Lee, and T. H. Brandon. 2019. Reasons for exclusion from a smoking cessation trial: An analysis by race/ethnicity. *Ethnicity & Disease* 29(1):23–30.
Hoppe, T. A., A. Litovitz, K. A. Willis, R. A. Meseroll, M. J. Perkins, B. I. Hutchins, A. F. Davis, M. S. Lauer, H. A. Valantine, J. M. Anderson, and G. M. Santangelo. 2019. Topic choice contributes to the lower rate of NIH awards to African-American/Black scientists. *Science Advances* 5(10):eaaw7238.
Hoyo, C., M. L. Reid, P. A. Godley, T. Parrish, L. Smith, and M. Gammon. 2003. Barriers and strategies for sustained participation of African-American men in cohort studies. *Ethnicity & Disease* 13(4):470–476.
HRSA (Health Resources and Services Administration). 2020a. National Health Center Program Uniform Data System (UDS) Awardee Data. https://data.hrsa.gov/tools/data-reporting/program-data/national.
HRSA. 2020b. National Health Center Program Uniform Data System (UDS) Look-Alike Data. https://data.hrsa.gov/tools/data-reporting/program-data/national-lookalikes.
HRSA. 2022. Health Center Data and Reporting. https://bphc.hrsa.gov/datareporting/index.html.
Huamani, K. F., B. Metch, G. Broder, M. Andrasik. A demographic analysis of racial/ethnic minority enrollment into HVTN preventive early phase HIV vaccine clinical trials conducted in the United States, 2002-2016. *2019 Public Health Reports* 134(1):72–80. doi: 10.1177/0033354918814260. Epub 2018 Dec 5. PMID: 30517057; PMCID: PMC6304725.
Huang, B., D. De Vore, C. Chirinos, J. Wolf, D. Low, R. Willard-Grace, S. Tsao, C. Garvey, D. Donesky, G. Su, and D. H. Thom. 2019. Strategies for recruitment and retention of underrepresented populations with chronic obstructive pulmonary disease for a clinical trial. *BMC Medical Research Methodology* 19(1).
Huang, H.-Y., M. O. Ezenwa, D. J. Wilkie, and M. K. Judge. 2013. ResearchTracking: Monitoring gender and ethnic minority recruitment and retention in cancer symptom studies. *Cancer Nursing* 36(3):E1–E6.
Hudson, K. L., M. S. Lauer, and F. S. Collins. 2016. Toward a new era of trust and transparency in clinical trials. *JAMA* 316(13):1353–1354.
Hudson, S. V., D. Memperousse, and H. Leventhal. 2005. Physician perspectives on cancer clinical trials and barriers to minority recruitment. *Cancer Control: Journal of the Moffitt Cancer Center* 12(Suppl. 2):93–96.
Hughes, T. B., V. R. Varma, C. Pettigrew, M. S. Albert, and B. J. Bowers. 2017. African Americans and clinical research: Evidence concerning barriers and facilitators to participation and recruitment recommendations. *Gerontologist* 57(2):348–358.
Hume, M., L. L. Lewis, and R. M. Nelson. 2017. Meeting the goal of concurrent adolescent and adult licensure of HIV prevention and treatment strategies. *Journal of Medical Ethics* 43(12):857–860.
Hurley, E. A. 2017. From the director: "Vulnerability" in the revised Common Rule. *Ampersand*, PRIMER (blog), posted September 12, 2017. https://blog.primr.org/vulnerability-revised-common-rule/.
IHS (Indian Health Service). 2021. Statement from the IHS acting director Elizabeth Fowler on the president's fiscal year 2022 budget. Press Release, May 28, 2021. https://www.ihs.gov/newsroom/pressreleases/2021-press-releases/statement-from-the-ihs-acting-director-elizabeth-fowler-on-the-presidents-fiscal-year-2022-budget/.

Iltis, A. S. 2004. Costs to subjects for research participation and the informed consent process: Regulatory and ethical considerations. *IRB: Ethics & Human Research* 26(6): 9–13. https://doi.org/10.2307/3564097.

Indorewalla, K. K., M. K. O'Connor, A. E. Budson, C. Guess DiTerlizzi, and J. Jackson. 2021. Modifiable barriers for recruitment and retention of older adults participants from underrepresented minorities in Alzheimer's disease research. *Journal of Alzheimer's Disease* 80(3):927–940.

IOM (Institute of Medicine). 1990. *Broadening the Base of Treatment for Alcohol Problems.* Washington, DC: The National Academies Press.

IOM. 1994. *Women and Health Research: Ethical and Legal Issues of Including Women in Clinical Studies, Volume 2, Workshop and Commissioned Papers,* A. C. Mastroianni, R. Faden, and D. Federman, eds. Washington, DC: The National Academies Press.

IOM. 2001. *Exploring the Biological Contributions to Human Health: Does Sex Matter?* T. M. Wizemann and M.-L. Pardue, eds. Washington, DC: The National Academies Press.

IOM. 2011. "Introduction," chap. 1 in *The Health of Lesbian, Gay, Bisexual and Transgender People: Building a Foundation for Better Understanding.* Washington, DC: The National Academies Press. https://www.nap.edu/read/13128/chapter/1.

IRS (Internal Revenue Service). n.d. Patient-Centered Outcomes Research Trust Fund Fee (IRC 4375, 4376 and 4377): Questions and Answers. https://www.irs.gov/affordable-care-act/patient-centered-outcomes-research-trust-fund-fee-questions-and-answers.

Ix, J. H., M. A. Allison, J. O. Denenberg, M. Cushman, and M. H. Criqui. 2008. Novel cardiovascular risk factors do not completely explain the higher prevalence of peripheral arterial disease among African Americans. The San Diego population study. *Journal of the American College of Cardiology* 51(24):2347–2354.

Jackson, J. L., C. Bates, S. M. Asch, R. Roberts, and J. R. Clarkson. 2021. How can medical journals promote equity and counter racism? *Journal of General Internal Medicine* 36(10):2919–2921.

Jacobs, E. A., I. Ganguli, and S. K. Inouye. 2021. Increasing representation of women as editors in medical journals. *JAMA Network Open* 4(9):e2123364–e2123364.

Jamerson, K., and V. DeQuattro. 1996. The impact of ethnicity on response to antihypertensive therapy. *American Journal of Medicine* 101(3A):22S–32S.

James, D. C. S., C. Harville II, O. Efunbumi, I. Babazadeh, and S. Ali. 2017. "You have to approach us right": A qualitative framework analysis for recruiting African Americans into mHealth research. *Health Education and Behavior* 44(5):781–790.

Jamison, D. T., L. H. Summers, G. Alleyne, K. J. Arrow, S. Berkley, A. Binagwaho, F. Bustreo, D. Evans, R. G. Feachem, J. Frenk, G. Ghosh, S. J. Goldie, Y. Guo, S. Gupta, R. Horton, M. E. Kruk, A. Mahmoud, L. K. Mohohlo, M. Ncube, A. Pablos-Mendez, K. S. Reddy, H. Saxenian, A. Soucat, K. H. Ulltveit-Moe, and G. Yamey. 2013. Global health 2035: A world converging within a generation. *Lancet* 382(9908):1898–1955.

Javid, S. H., J. M. Unger, J. R. Gralow, C. M. Moinpour, A. J. Wozniak, J. W. Goodwin, P. N. Lara, P. A. Williams, L. F. Hutchins, C. C. Gotay, and K. S. Albain. 2012. A prospective analysis of the influence of older age on physician and patient decision-making when considering enrollment in breast cancer clinical trials (SWOG S0316). *Oncologist* 17(9):1180–1190.

Jones, N. C., A. K. Otto, D. E. Ketcher, J. B. Permuth, G. P. Quinn, and M. B. Schabath. 2020. Inclusion of transgender and gender diverse health data in cancer biorepositories. *Contemporary Clinical Trials Communications* 19(10).

Jonsen, A. R. 1978. Research involving children: Recommendations of the national commission for the protection of human subjects of biomedical and behavioral research. *Pediatrics* 62(2):131–136.

Joosten, Y. A., T. L. Israel, N. A. Williams, L. R. Boone, D. G. Schlundt, C. P. Mouton, R. S. Dittus, G. R. Bernard, and C. H. Wilkins. 2015. Community engagement studios: A structured approach to obtaining meaningful input from stakeholders to inform research. *Academic Medicine* 90(12):1646–1650.

Joseph, G., and D. Dohan. 2009. Diversity of participants in clinical trials in an academic medical center: The role of the 'good study patient?'. *Cancer* 115(3):608–615.

Kalliainen, L. K., I. Wisecarver, A. Cummings, and J. Stone. 2018. Sex bias in hand surgery research. *The Journal of Hand Surgery* 43(11):1026–1029.
Kalt, J. P., and J. W. Singer. 2004. *Myths and Realities of Tribal Sovereignty: The Law and Economics of Indian Self-rule*. Faculty Research Working Paper Series. John F. Kennedy School of Government, Harvard University. https://scholar.harvard.edu/files/jsinger/files/myths_realities.pdf.
Ka'opua, L. S., D. Mitschke, and J. Lono. 2004. Increasing participation in cancer research: Insights from Native Hawaiian women in medically underserved communities. *Pacific Health Dialog* 11(2):170–175.
Kaplan, C. P., A. M. Nápoles, S. Narine, S. Gregorich, J. Livaudais-Toman, T. Nguyen, Y. Leykin, M. Roach, and E. J. Small. 2015. Knowledge and attitudes regarding clinical trials and willingness to participate among prostate cancer patients. *Contemporary Clinical Trials* 45:443–448.
Karris, M. Y., K. Dubé, and A. A. Moore. 2020. What lessons it might teach us? Community engagement in HIV research. *Current Opinion in HIV and AIDS* 15(2):142–149. https://doi.org/10.1097/COH.0000000000000605.
Kass, N. E., L. Chaisson, H. A. Taylor, and J. Lohse. 2011. Length and complexity of US and international HIV consent forms from federal HIV network trials. *Journal of General Internal Medicine* 26(11):1324–1328.
Katz, R. V., S. S. Kegeles, N. R. Kressin, B. L. Green, S. A. James, M. Q. Wang, S. L. Russell, and C. Claudio. 2008. Awareness of the Tuskegee Syphilis Study and the US presidential apology and their influence on minority participation in biomedical research. *American Journal of Public Health* 98(6):1137–1142.
Katz, R. V., S. S. Kegeles, N. R. Kressin, B. L. Green, M. Q. Wang, S. A. James, S. L. Russell, and C. Claudio. 2006. The Tuskegee Legacy Project: Willingness of minorities to participate in biomedical research. *Journal of Health Care for the Poor and Underserved* 17(4):698–715.
Kessler Foundation. 2011. Kessler foundation receives grants to study health disparities in people with disabilities. Press Release, November 26, 2011. https://kesslerfoundation.org/press-releases/kessler-foundation-receives-grants-study-health-disparities-people-disabilities.
KFF (Kaiser Family Foundation). 2019. Poverty Rate by Race/Ethnicity. https://www.kff.org/other/state-indicator/poverty-rate-by-raceethnicity/?currentTimeframe=0&sortModel=%7B%22colId%22:%22Location%22,%22sort%22:%22asc%22%7D.
Khan, M., I. Shahid, T. Siddiqi, S. Khan, H. Warraich, S. Greene, J. Butler, and E. Michos. 2020. Ten-year trends in enrollment of women and minorities in pivotal trials supporting recent US Food and Drug Administration approval of novel cardiometabolic drugs. *Journal of the American Heart Association* 9(11):e015594.
Khan, S. U., M. Z. Khan, A. N. Lone, C. R. Subramanian, S. Talluri, J. K. Han, N. Isakadze, A. S. Volgman, and E. D. Michos. 2020. Abstract 14171: Association of women authors with women enrollment in the clinical trials of atrial fibrillation. *Circulation* 142(Suppl. 3).
Kibbe, M. R., and J. Freischlag. 2020. Call to action to all surgery journal editors for diversity in the editorial and peer review process. *JAMA Surgery* 155(11):1015–1016.
Kim, E. S., T. P. Carrigan, and V. Menon. 2008. Enrollment of women in National Heart, Lung, and Blood Institute-funded cardiovascular randomized controlled trials fails to meet current federal mandates for inclusion. *Journal of the American College of Cardiology* 52(8):672–673.
Kimminau, K. S., C. Jernigan, J. LeMaster, L. S. Aaronson, M. Christopher, S. Ahmed, A. Boivin, M. DeFino, R. Greenlee, G. Salvalaggio, D. Hendricks, C. Herbert, N. M. Mabachi, A. Macaulay, J. M. Westfall, and L. R. Waitman. 2018. Patient vs. Community Engagement: Emerging Issues. *Medical Care* 10 (Suppl. 1):S53–S57.
King, A. C., D. Cao, C. C. Southard, and A. Matthews. 2011. Racial differences in eligibility and enrollment in a smoking cessation clinical trial. *Health Psychology* 30(1):40–48.
King, W. D., D. Defreitas, K. Smith, J. Andersen, L. P. Perry, T. Adeyemi, J. Mitty, J. Fritsche, C. Jeffries, M. Littles, M. Fischl, G. Pavlov, and D. Mildvan. 2007. Attitudes and perceptions of AIDS clinical trials group site coordinators on HIV clinical trial recruitment and retention: A descriptive study. *AIDS Patient Care and STDs* 21(8):551–563.

King, J. B., L. C. Pinheiro, J. B.Ringel, A. P. Bress, D. Shimbo, P. Mutner, K. Reynolds, M. Cushman, G. Howard, J. J. Manly, and M. M. Safford. 2021. Multiple social vulnerabilities to health disparities and hypertension and death in the regards study. *Hypertension* 79(1):196–206.

Kneipp, S. M., T. A. Schwartz, D. J. Drevdahl, M. K. Canales, S. Santacroce, H. P. Santos Jr., and R. Anderson. 2018. Trends in health disparities, health inequity, and social determinants of health research: A 17-year analysis of NINR, NCI, NHLBE, and NIMHD funding. *Nursing Research* 67(3):231–241.

Knopf, A., M. A. Ott, C. B. Draucker, J. D. Fortenberry, D. H. Reirden, R. Arrington-Sanders, J. Schneider, D. Straub, R. Baker, G. Bakoyannis, and G. D. Zimet. 2020. Innovative approaches to obtain minors' consent for biomedical HIV prevention trials: Multi-site quasi-experimental study of adolescent and parent perspectives. *JMIR Research Protocols* 9(3):e16509.

Knopf A. S., M. A. Ott, N. Liu, B. G. Kapogiannis, G. D. Zimet, J. D. Fortenberry, and S. G. Hosek. 2017. Minors' and young adults' experiences of the research consent process in a phase II safety study of pre-exposure prophylaxis for HIV. *Journal of Adolescent Health* 61(6):747–754.

Knowler, W. C., E. Barrett-Connor, S. E. Fowler, R. F. Hamman, J. M. Lachin, E. A. Walker, and D. M. Nathan. 2002. Reduction in the incidence of type 2 diabetes with lifestyle intervention or metformin. *New England Journal of Medicine* 346(6):393–403.

Kopelman, L. M. 2000. Children as research subjects: A dilemma. *The Journal of Medicine and Philosophy* 25(6):745–764.

Kornstein, S. G., A. F. Schatzberg, M. E. Thase, K. A. Yonkers, J. P. McCullough, G. I. Keitner, A. J. Gelenberg, S. M. Davis, W. M. Harrison, and M. B. Keller. 2000. Gender differences in treatment response to sertraline versus imipramine in chronic depression. *American Journal of Psychiatry* 157(9):1445–1452.

Kost, R. G., A. Leinberger-Jabari, T. H. Evering, P. R. Holt, M. Neville-Williams, K. S. Vasquez, B. S. Coller, and J. N. Tobin. 2017. Helping basic scientists engage with community partners to enrich and accelerate translational research. *Academic Medicine* 92(3):374–379.

Kretzmann, J. P., and J. L. McKnight. 1993. *Building Communities from the Inside Out : A Path Toward Finding and Mobilizing a Community's Assets*. The Asset-Based Community Development Institute. Chicago: DePaul University Steans Center. https://resources.depaul.edu/abcd-institute/publications/Documents/Chapter 5 - PROVIDING SUPPORT FOR ASSET-BASED DEVELOPMENT- POLICIES AND GUIDELINES.pdf.

Krieger, N., and S. Sidney. 1996. Racial discrimination and blood pressure; The CARDIA study of young black and white adults. *American Journal of Public Health* 86: 1370-1378.

Kripalani, S., W. J. Heerman, N. J. Patel, N. Jackson, K. Goggins, R. L. Rothman, V. M. Yeh, K. A. Wallston, D. T. Smoot, and C. H. Wilkins. 2019. Association of health literacy and numeracy with interest in research participation. *Journal of General Internal Medicine* 34(4):544–551.

Kuhn, N. S., M. Parker, and C. Lefthand-Begay. 2020. Indigenous research ethics requirements: An examination of six tribal institutional review board applications and processes in the United States. *Journal of Empirical Research on Human Research Ethics* 15(4):279–291. doi:10.1177/1556264620912103.

Lakdawalla, D. N., D. P. Goldman, and B. Shang. 2005. The health and cost consequences of obesity among the future elderly. *Health Affairs* 24(Suppl. 2):W5-R30–W35-R41.

Langford, A. T., K. Resnicow, E. P. Dimond, A. M. Denicoff, D. S. Germain, W. McCaskill-Stevens, R. A. Enos, A. Carrigan, K. Wilkinson, and R. S. Go. 2014. Racial/ethnic differences in clinical trial enrollment, refusal rates, ineligibility, and reasons for decline among patients at sites in the National Cancer Institute's Community Cancer Centers Program. *Cancer* 120(6):877–884.

Largent, E. A., C. Grady, F. G. Miller, and A. Wertheimer. 2012. Money, coercion, and undue inducement: Attitudes about payments to research participants. *IRB: Ethics & Human Research* 34(1):1–8.

Largent, E. A., and H. F. Lynch. 2017. Paying research participants: The outsized influence of "undue influence." *IRB: Ethics & Human Research* 39(4):1–9.

Larkey, L. K., J. A. Gonzalez, L. E. Mar, and N. Glantz. 2009. Latina recruitment for cancer prevention education via community based participatory research strategies. *Contemporary Clinical Trials* 30(1):47–54.

Las Nueces, D., K. Hacker, A. Digirolamo, and L. S. Hicks. 2012. A systematic review of community-based participatory research to enhance clinical trials in racial and ethnic minority groups. *Health Services Research* 47(3pt2):1363–1386.

Lauer, M. 2020. Some thoughts following the NIH inclusion across the lifespan 2 workshop. National Institutes of Health, Extramural Nexus (blog), posted December 10, 2020. https://nexus.od.nih.gov/all/2020/12/10/some-thoughts-following-the-nih-inclusion-across-the-lifespan-2-workshop/.

Lawrence, J. 2000. The Indian Health Service and the sterilization of Native American women. *American Indian Quarterly* 24(3):400–419.

Leaf, D. E., B. Tysinger, D. P. Goldman, and D. N. Lakdawalla. 2020. Predicting quantity and quality of life with the future elderly model. *Health Economics*. Special Issue Paper, Wiley Online Library. https://onlinelibrary.wiley.com/doi/abs/10.1002/hec.4169.

Levitan, B., K. Getz, E. L. Eisenstein, M. Goldberg, M. Harker, S. Hesterlee, B. Patrick-Lake, J. N. Roberts, and J. Dimasi. 2018. Assessing the financial value of patient engagement: A quantitative approach from CTTI's patient groups and clinical trials project. *Therapeutic Innovation & Regulatory Science* 52(2):220–229.

Lewis, J. H., M. L. Kilgore, D. P. Goldman, E. L. Trimble, R. Kaplan, M. J. Montello, M. G. Housman, and J. J. Escarce. 2003. Participation of patients 65 years of age or older in cancer clinical trials. *Journal of Clinical Oncology* 21(7):1383–1389.

Lewison, D. 1997. *Retailing*. Hoboken, NJ: Prentice Hall.

Liu, K. A., and N. A. Dipietro Mager. 2016. Women's involvement in clinical trials: Historical perspective and future implications. *Pharmacy Practice (Granada)* 14(1):708.

Lockett, J. 2017. Let's talk about inclusion of all ages in research. National Institute on Aging, Research and Funding (blog), posted April 26, 2017. https://www.nia.nih.gov/research/blog/2017/04/lets-talk-about-inclusion-all-ages-research.

Lor, M., and B. J. Bowers. 2018. Hmong older adults' perceptions of insider and outsider researchers: Does it matter for research participation? *Nursing Research* 67(3):222–230.

Lowsky, D. J., S. J. Olshansky, J. Bhattacharya, and D. P. Goldman. 2014. Heterogeneity in healthy aging. *Journals of Gerontology, Series A: Biological Sciences and Medical Sciences* 69(6):640–649.

Lucero, J. E., B. Boursaw, M. M. Eder, E. Greene-Moton, N. Wallerstein, and J. G. Oetzel. 2020. Engage for equity: The role of trust and synergy in community-based participatory research. *Health Education & Behavior* 47(3):372–379.

Ludmir, E. B., W. Mainwaring, T. A. Lin, A. B. Miller, A. Jethanandani, A. F. Espinoza, J. J. Mandel, S. H. Lin, B. D. Smith, G. L. Smith, N. A. VanderWalde, B. D. Minsky, A. C. Koong, T. E. Stinchcombe, R. Jagsi, D. R. Gomez, C. R. Thomas Jr., and C. D. Fuller. 2019. Factors associated with age disparities among cancer clinical trial participants. *JAMA Oncology* 5(12):1769–1773.

MacLeod, S. 2010. Therapeutic drug monitoring in pediatrics: How do children differ? *Therapeutic Drug Monitoring* 32(3):253–256.

Maixner, S. M., A. M. Mellow, and R. Tandon. 1999. The efficacy, safety, and tolerability of antipsychotics in the elderly. *Journal of Clinical Psychiatry* 60 (Suppl. 8):29–41.

Mak, W. W. S., R. W. Law, J. Alvidrez, and E. J. Pérez-Stable. 2007. Gender and ethnic diversity in NIMH-funded clinical trials: Review of a decade of published research. *Administration and Policy in Mental Health and Mental Health Services Research* 34(6):497–503.

Malzbender, K., L. Lavin-Mena, L. Hughes, N. Bose, D. Goldman, D. Patel. 2020. Key barriers to clinical trials for Alzheimer's disease. White Paper. Leonard D. Schaeffer Center for Health Policy and Economics. Los Angeles: University of Southern California.

Manders, D. B., A. Paulsen, D. L. Richardson, S. M. Kehoe, D. S. Miller, and J. S. Lea. 2014. Factors associated with clinical trial screening failures in gynecologic oncology. *Gynecologic Oncology* 134(3):450–454.

Martin, A. R., M. Kanai, Y. Kamatani, Y. Okada, B. M. Neale, and M. J. Daly. 2019. Clinical use of current polygenic risk scores may exacerbate health disparities. *Nature Genetics* 51(4):584–591.

Maxwell, A. E., R. Bastani, P. Vida, and U. S. Warda. 2005. Strategies to recruit and retain older Filipino-American immigrants for a cancer screening study. *Journal of Community Health* 30(3):167–179.

McCarthy-Keith, D., S. Nurudeen, A. Armstrong, E. Levens, and L. Nieman. 2010. Recruitment and retention of women for clinical leiomyoma trials. *Contemporary Clinical Trials* 31(1).

McCormick, R. A. 1974. Proxy consent in the experimentation situation. Pp. 297–309 in *Biomedical Ethics and the Law*, J. M. Humber and R. F. Almeder, eds. Boston: Springer.

McElfish, P. A., C. R. Long, J. P. Selig, B. Rowland, R. S. Purvis, L. James, A. Holland, H. C. Felix, and M. R. Narcisse. 2018. Health research participation, opportunity, and willingness among minority and rural communities of Arkansas. *Clinical and Translational Science* 11(5):487–497.

McKee, M., D. Schlehofer, and D. Thew. 2013. Ethical issues in conducting research with deaf populations. *American Journal of Public Health* 103(12):2174–2178.

McMurdo, M. E. T., H. Roberts, S. Parker, N. Wyatt, H. May, C. Goodman, S. Jackson, J. Gladman, S. O'Mahony, K. Ali, E. Dickinson, P. Edison, and C. Dyer. 2011. Improving recruitment of older people to research through good practice. *Age and Ageing* 40(6):659–665.

Medicare advantage: Reforms needed to ensure access to clinical trials. 2009. *Journal of Oncology Practice* 5(3):144–145.

Meeker-O'Connell, A., and J. Menikoff. 2021. *The role of institutional review boards*: National Academies of Sciences, Engineering, and Medicine. https://www.nationalacademies.org/documents/embed/link/LF2255DA3DD1C41C0A42D3BEF0989ACAECE3053A6A9B/file/D3A0C8EA19ED6783B1E59A8D5ED5A7607F3E15C782BC.

Michos, E. D., T. K. Reddy, M. Gulati, L. C. Brewer, R. M. Bond, G. P. Velarde, A. L. Bailey, M. R. Echols, S. A. Nasser, H. E. Bays, A. M. Navar, and K. C. Ferdinand. 2021. Improving the enrollment of women and racially/ethnically diverse populations in cardiovascular clinical trials: An ASPC practice statement. *American Journal of Preventive Cardiology* 8:100250.

Michos, E. D., and H. G. C. Van Spall. 2021. Increasing representation and diversity in cardiovascular clinical trial populations. *Nature Reviews Cardiology* 18(8):537–538.

Mistretta, A. 2016. NIH outreach toolkit: How to engage, recruit, and retain women in clinical research. Paper presented at APHA 2016 Annual Meeting & Expo (October 29–November 2, 2016). https://orwh.od.nih.gov/toolkit.

Moher, D., S. Hopewell, K. F. Schulz, V. Montori, P. C. Gøtzsche, P. J. Devereaux, D. Elbourne, M. Egger, and D. G. Altman. 2010. CONSORT 2010 explanation and elaboration: Updated guidelines for reporting parallel group randomised trials. *BMJ* 340:c869.

Moher, D., S. Hopewell, K. Schulz, V. Montori, P. Gøtzsche, P. Devereaux, D. Elbourne, M. Egger, and D. Altman. 2012. CONSORT 2010 explanation and elaboration: Updated guidelines for reporting parallel group randomised trials. *International Journal of Surgery* 10(1):28–55.

Moreno-John, G., A. Gachie, C. M. Fleming, A. Napoles-Springer, E. Mutran, S. M. Manson, and E. J. Perez-Stable. 2004. Ethnic minority older adults participating in clinical research: Developing trust. *Journal of Aging and Health* 16(5 Suppl.):93S–123S.

Murphy, E., and A. Thompson. 2009. An exploration of attitudes among Black Americans towards psychiatric genetic research. *Psychiatry* 72(2):177–194.

Murphy, E. J., P. Wickramaratne, and M. M. Weissman. 2009. Racial and ethnic differences in willingness to participate in psychiatric genetic research. *Psychiatric Genetics* 19(4):186–194.

Murthy, V. H., H. M. Krumholz, and C. P. Gross. 2004. Participation in cancer clinical trials: Race-, sex-, and age-based disparities. *JAMA* 291(22):2720–2726.

Nadel, M. V. 1990. "National Institutes of Health: Problems in Implementing Policy on Women in Study Populations." Statement of Mark V. Nadel, Associate Director, National and Public Health Issues, Human Resources Division, before the Subcommittee on Housing and Consumer Interest, Select Committee on Aging, U.S. House of Representatives. Released July 24, 1990. Washington, DC: U.S. General Accounting Office. http://archive.gao.gov/d48t13/141859.pdf.

REFERENCES

Nanna, M. G., S. T. Chen, A. J. Nelson, A. M. Navar, and E. D. Peterson. 2020. Representation of older adults in cardiovascular disease trials since the inclusion across the lifespan policy. *JAMA Internal Medicine* 180(11):1531–1533.

NARHC (National Association of Rural Health Clinics). 2022. Mission Statement. https://www.narhc.org/narhc/Mission_Statement.asp.

NASEM (National Academies of Sciences, Engineering, and Medicine). 2015. *The Growing Gap in Life Expectancy by Income: Implications for Federal Programs and Policy Responses*. Washington, DC: The National Academies Press.

NASEM. 2018. *The Next Generation of Biomedical and Behavioral Sciences Researchers: Breaking through*, R. Daniels and L. Beninson, eds. Washington, DC: The National Academies Press.

NASEM. 2020a. *Promising Practices for Addressing the Underrepresentation of Women in Science, Engineering, and Medicine: Opening Doors*, R. Colwell, A. Bear, and A. Helman, eds. Washington, DC: The National Academies Press.

NASEM. 2020b. *Understanding the Well-Being of LGBTQI+ Populations*, M. J. Sepulveda, C. J. Patterson, and J. White, eds. Washington, DC: The National Academies Press.

Nature Medicine. 2018. Diversifying clinical trials. 24(1779). https://doi.org/10.1038/s41591-018-0303-4.

Nazha, B., M. Mishra, R. Pentz, and T. Owonikoko. 2019. Enrollment of racial minorities in clinical trials: Old problem assumes new urgency in the age of immunotherapy. *American Society of Clinical Oncology Educational Book*, Vol. 39. Alexandria, VA: American Society of Clinical Oncology.

National Commission for the Protection of Human Subjects of Biomedical and Behavioral Research. (1979). The Belmont report: Ethical principles and guidelines for the protection of human subjects of research. U.S. Department of Health and Human Services. https://www.hhs.gov/ohrp/regulations-and-policy/belmont-report/read-the-belmont-report/index.html.

NCAI (National Congress of American Indians). 2016. *Reducing Disparities in the Federal Healthcare Budget*. https://www.ncai.org/resources/ncai-publications/08_FY2017_health_care.pdf.

NCPHS. 1978. *Research Involving Children: Report and Recommendations of the National Commission for the Protection of Human Subjects of Biomedical and Behavioral Research*. (45 C.F.R. 46.). *Federal Register* 43:2083.

NCUIH (National Council of Urban Indian Health). 2019. Advance Appropriations: Protecting Tribal Communities from the Effects of Government Shutdowns Hearing, House Committee on Natural Resources Subcommittee for Indigenous Peoples of the United States. Statement of Maureen Rosette, President, NCUIH, September 25, 2019. https://docs.house.gov/meetings/II/II24/20190925/110050/HHRG-116-II24-20190925-SD013.pdf.

NEJM. 2021. Striving for diversity in research studies. Editorial, October 7, 2021. *New England Journal of Medicine* 385(15):1429–1430.

Nelson, L. A., S. E. Williamson, L. M. LeStourgeon, and L. S. Mayberry. 2021. Retaining diverse adults with diabetes in a long-term trial: Strategies, successes, and lessons learned. *Contemporary Clinical Trials* 105:106388.

Newman, P. A., N. Duan, K. J. Roberts, D. Seiden, E. T. Rudy, D. Swendeman, and S. Popova. 2006. HIV vaccine trial participation among ethnic minority communities: Barriers, motivators, and implications for recruitment. *Journal of Acquired Immune Deficiency Syndromes* 41(2):210–217.

Nguyen, T., N. Wallerstein, R. Das, M. Sabado-Liwag, B. Bird, T. Jacob, T. Cannady, L. Martinez, U. Ndulue, A. Ortiz, A. Stubbs, L. Pichon, S. Tanjarsiri, J. Pang, and K. Woo. 2021. Conducting community-based participatory research with minority communities to reduce health disparities. Pp. 171–186 in *The Science of Health Disparities Research*, I. Dankwa-Mullan, E. J. Pérez-Stable, K. L. Gardner, X. Zhang, and A. M. Rosariok, eds. Hoboken, NJ: John Wiley & Sons, Inc.

Nguyen, T. T., C. P. Somkin, Y. Ma, L. C. Fung, and T. Nguyen. 2005. Participation of Asian-American women in cancer treatment research: A pilot study. *Journal of the National Cancer Institute* (Monograph) (35):102–105.

Nielsen, M. W., J. P. Andersen, L. Schiebinger, and J. W. Schneider. 2017. One and a half million medical papers reveal a link between author gender and attention to gender and sex analysis. *Nature Human Behaviour* 1(11):791–796.

Night, S. 2009. History of NIH Revitalization Act of 1993 and current relevance to inclusion of women and minorities in clinical research. *SoCRA Source* August 2009:20–23. https://nebula.wsimg.com/3809b19c768db79eb0d57d86d4940699?AccessKeyId=4ECD43F4A65F6DBF7F21&disposition=0&alloworigin=1.

NIH. 1987. *NIH Guide for Grants and Contracts.* https://grants.nih.gov/grants/guide/historical/1987_01_09_Vol_16_No_01.pdf.

NIH. 1989. *NIH Guide for Grants and Contracts.* https://grants.nih.gov/grants/guide/historical/1989_08_11_Vol_18_No_27.pdf.

NIH. 1998. *NIH Policy and Guidelines on the Inclusion of Children as Participants in Research Involving Human Subjects.* https://grants.nih.gov/grants/guide/notice-files/not98-024.html.

NIH. 2001a. *NIH Policy and Guidelines on the Inclusion of Women and Minorities as Subjects in Clinical Research.* https://grants.nih.gov/policy/inclusion/women-and-minorities/guidelines.htm.

NIH. 2001b. *Amendment: NIH Policy and Guidelines on the Inclusion of Women and Minorities as Subjects in Clinical Research.* https://grants.nih.gov/grants/guide/notice-files/NOT-OD-02-001.html.

NIH. 2010. NIH announces institute on minority health and health disparities. News Release, September 27, 2010. https://www.nih.gov/news-events/news-releases/nih-announces-institute-minority-health-health-disparities.

NIH. 2013. *Draft Report of the Advisory Committee to the Director Working Group on Diversity in the Biomedical Research Workforce.* https://www.acd.od.nih.gov/documents/reports/DiversityBiomedicalResearchWorkforceReport.pdf.

NIH. 2015a. *National Institutes of Health FY2012 Sexual and Gender Minority Health Research Portfolio Analysis Report.* https://dpcpsi.nih.gov/sgmro/reports.

NIH. 2015b. NIH Precision Medicine Initiative. https://www.nimhd.nih.gov/about/legislative-info/clips/pmi.html.

NIH. 2015c. Request for information: NIH precision medicine cohort. Notice Number: NOT-OD-15-096, Release Date, April 20, 2015. http://grants.nih.gov/grants/guide/notice-files/NOT-OD-15-096.html.

NIH. 2016. Era Commons – User Registration. https://grants.nih.gov/grants/how-to-apply-application-guide/prepare-to-apply-and-register/registration/investigators-and-other-users/era-commons-user-registration.htm.

NIH. 2017a. *Amendment: NIH Policy and Guidelines on the Inclusion of Women and Minorities as Subjects in Clinical Research*, Notice Number NOT-OD-18-014. https://grants.nih.gov/grants/guide/notice-files/NOT-OD-18-014.html.

NIH. 2017b. *Revision: NIH Policy and Guidelines on the Inclusion of Individuals across the Lifespan as Participants in Research Involving Human Subjects*, Notice Number NOT-OD-18-116. https://grants.nih.gov/grants/guide/notice-files/NOT-OD-18-116.html.

NIH. 2017c. Information about NIH Clinical Trial Stewardship. https://grants.nih.gov/policy/clinical-trials/why-changes.htm.

NIH. 2018a. Post-Award Monitoring and Reporting. https://grants.nih.gov/grants/post-award-monitoring-and-reporting.htm.

REFERENCES

NIH. 2018b. *Task Force on Research Specific to Pregnant Women and Lactating Women: Report to Secretary, Health and Human Services and Congress.* https://www.nichd.nih.gov/sites/default/files/2018-09/PRGLAC_Report.pdf.

NIH. 2019a. Task Force on Research Specific to Pregnant Women and Lactating Women (PRGLAC). https://www.nichd.nih.gov/about/advisory/PRGLAC.

NIH. 2019b. Guidelines for the review of inclusion on the basis of sex/gender, race, ethnicity, and age in clinical research. https://grants.nih.gov/grants/peer/guidelines_general/Review_Human_subjects_Inclusion.pdf.

NIH. 2020a. *NIH Inclusion Across the Lifespan II Workshhop Report.* https://grants.nih.gov/sites/default/files/IAL-II-Workshop-Report.pdf.

NIH. 2020b. Center for Scientific Review. https://public.csr.nih.gov/

NIH. 2021a. *FDAAA 801 and the Final Rule.* https://clinicaltrials.gov/ct2/manage-recs/fdaaa.

NIH. 2021b. *Strategic Plan to Advance Research on the Health and Well-being of Sexual & Gender Minorities: Fiscal Years 2021–2025.* https://dpcpsi.nih.gov/sites/default/files/SGMStrategic-Plan_2021_2025.pdf.

NIH. 2021c. Faculty Institutional Recrtuiment for Sustainable Transformation (FIRST). https://commonfund.nih.gov/first/programhighlights.

NIH. 2021d. FAQs About Research Supplements to Promote Diversity in Health-Related Research (Diversity Supplements). https://nida.nih.gov/about-nida/organization/offices/office-nida-director-od/office-research-training-diversity-disparities-ortdd/odhd/faqs-about-research-supplements-to-promote-diversity-in-health-related.

NIH. 2022. *NIH RCDC Inclusion Statistics Report.* https://report.nih.gov/risr/#/.

NIHB (National Indian Health Board). 2020. *Reclaiming Tribal Health: A National Budget Plan to Rise above Failed Policies and Fufill Trust Obligation to Tribal Nations.* https://www.nihb.org/communications/nihb_publications.php.

NIMHD (National Institute of Minority Health and Health Disparities). 2022. History. https://www.nimhd.nih.gov/about/overview/history/.

Nipp, R. D., K. Hong, and E. D. Paskett. 2019. Overcoming barriers to clinical trial enrollment. *American Society of Clinical Oncology Educational Book* (39):105–114.

Nipp, R. D., H. Lee, E. Powell, N. E. Birrer, E. Poles, D. Finkelstein, K. Winkfield, S. Percac-Lima, B. Chabner, and B. Moy. 2016. Financial burden of cancer clinical trial participation and the impact of a cancer care equity program. *Oncologist* 21(4):467–474.

Nours, S. 2021. Improving the representation of women and underrepresented minorities in clinical trials and research. In *NASEM Report Interviews*, L. Bothwell, ed. https://nap.nationalacademies.org/resource/26479/Bothwell_Assessing_Federal_Policies_on_the_Inclusion_of_Women_and_Minorities-Clinical_Trials.pdf.

NRC (National Research Council). 2005. *Assessment of NIH Minority Research and Training Programs: Phase 3.* Washington, DC: The National Academies Press.

NSF (National Science Foundation). 2021. *Women, Minorities, and Persons with Disabilities in Science and Engineering.* National Center for Science and Engineering Statistics. Arlington, VA: National Science Foundation.

O'Brien, E. C., S. R. Raman, A. Ellis, B. G. Hammill, L. G. Berdan, T. Rorick, S. Janmohamed, Z. Lampron, A. F. Hernandez, and L. H. Curtis. 2021. The use of electronic health records for recruitment in clinical trials: A mixed methods analysis of the Harmony Outcomes Electronic Health Record Ancillary Study. *Trials* 22(1):465.

Occa, A., S. E. Morgan, and J. N. E. Potter. 2018. Underrepresentation of Hispanics and other minorities in clinical trials: Recruiters' perspectives. *Journal of Racial and Ethnic Health Disparities* 5(2):322–332.

Odierna, D., and L. Bero. 2009. Systematic reviews reveal unrepresentative evidence for the development of drug formularies for poor and nonwhite populations. *Journal of Clinical Epidemiology* 62(12):1268–1278.

Ofili, E. O., L. E. Schanberg, B. Hutchinson, F. Sogade, I. Fergus, P. Duncan, J. Hargrove, A. Artis, O. Onyekwere, W. Batchelor, M. Williams, A. Oduwole, A. Onwuanyi, F. Ojutalayo, J. A. Cross, T. B. Seto, H. Okafor, P. Pemu, L. Immergluck, M. Foreman, E. A. Mensah, A. Quarshie, M. Mubasher, A. Baker, A. Ngare, A. Dent, M. Malouhi, P. Tchounwou, J. Lee, T. Hayes, M. Abdelrahim, D. Sarpong, E. Fernandez-Repollet, S. O. Sodeke, A. Hernandez, K. Thomas, A. Dennos, D. Smith, D. Gbadebo, J. Ajuluchikwu, B. W. Kong, C. McCollough, S. R. Weiler, M. D. Natter, K. D. Mandl, and S. Murphy. 2019. The Association of Black Cardiologists (ABC) Cardiovascular Implementation Study (CVIS): A research registry integrating social determinants to support care for underserved patients. *International Journal of Environmental Research and Public Health* 16(9):1631.

Oh, S. S., J. Galanter, N. Thakur, M. Pino-Yanes, N. E. Barcelo, M. J. White, D. M. de Bruin, R. M. Greenblatt, K. Bibbins-Domingo, and A. H. Wu. 2015. Diversity in clinical and biomedical research: A promise yet to be fulfilled. *PLOS Medicine* 12(12):e1001918.

Olin, J. T., K. S. Dagerman, L. S. Fox, B. Bowers, and L. S. Schneider. 2002. Increasing ethnic minority participation in Alzheimer disease research. *Alzheimer Disease and Associated Disorders* 16:A82–S85.

Olshansky, S. J., T. Antonucci, L. Berkman, R. H. Binstock, A. Boersch-Supan, J. T. Cacioppo, B. A. Carnes, L. L. Carstensen, L. P. Fried, D. P. Goldman, J. Jackson, M. Kohli, J. Rother, Y. Zheng, and J. Rowe. 2012. Differences in life expectancy due to race and educational differences are widening, and many may not catch up. *Health Affairs (Millwood)* 31(8):1803–1813.

Olshansky, S. J., D. P. Goldman, Y. Zheng, and J. W. Rowe. 2009. Aging in america in the twenty-first century: Demographic forecasts from the MacArthur Foundation Research Network on an Aging Society. *Milbank Q* 87(4):842–862.

O'Mara-Eves, A., G. Brunton, D. McDaid, S. Oliver, J. Kavanagh, F. Jamal, T. Matosevic, A. Harden, and J. Thomas. 2013. Community engagement to reduce inequalities in health: A systematic review, meta-analysis and economic analysis. *Public Health Research* 1(4):1–526.

Ong, M. 2021. Single digits: Black, Hispanic scientists strikingly underrepresented at NCI among senior workforce and grantees. *The Cancer Letter* 47(26).

Ortman, J. M., and C. E. Guarneri. 2009. United States population projections: 2000 to 2050. *United States Census Bureau* 1(15):1–19.

Otado, J., J. Kwagyan, D. Edwards, A. Ukaegbu, F. Rockcliffe, and N. Osafo. 2015. Culturally competent strategies for recruitment and retention of African American populations into clinical trials. *Clinical and Translational Science* 8(5):460–466.

Paskett, E. D., M. R. Cooper, N. Stark, T. C. Ricketts, S. Tropman, T. Hatzell, T. Aldrich, and J. Atkins. 2002. Clinical trial enrollment of rural patients with cancer. *Cancer Practice* 10(1):28–35.

Passmore, S. R., D. Farrar Edwards, C. A. Sorkness, S. Esmond, and A. R. Brasier. 2020. Training needs of investigators and research team members to improve inclusivity in clinical and translational research participation. *Journal of Clinical and Translational Science* 5(1):e57.

Patterson, C. H. 1985. Respect (unconditional positive regard). Pp. 59–63 in *The Therapeutic Relationship*. Monterey, CA: Brooks/Cole.

PCORI (Patient-Centered Outcomes Research Institute). n.d. *Diversity and Inclusion in PCORnet: Need and Recommendations.* https://www.pcori.org/sites/default/files/3928-PCORnet-Engagement-Committee-Diversity-Inclusion-Needs-Recommendations.pdf.

PCORI. 2019. *PRIDEnet: A Participant-Powered Research Network of Sexual and Gender Minorities.* https://www.pcori.org/research-results/2015/pridenet-participant-powered-research-network-sexual-and-gender-minorities.

PCORI. 2020. *2020 Annual Report.* Washington DC: Patient-Centered Outcomes Research Institute.

Penberthy, L. T., B. A. Dahman, V. I. Petkov, and J. P. DeShazo. 2012. Effort required in eligibility screening for clinical trials. *Journal of Oncology Practice* 8(6):365–370.

Pinho-Gomes, A.-C., A. Vassallo, K. Thompson, K. Womersley, R. Norton, and M. Woodward. 2021. Representation of women among editors in chief of leading medical journals. *JAMA Network Open* 4(9):e2123026–e2123026.

Pinn, V. W. 2003. Sex and gender factors in medical studies: Implications for health and clinical practice. *JAMA* 289(4):397–400.
Pearson, C. Schapiro, L., Pearson, S.D. 2022. The next generation of rare disease drug policy: Ensuring both innovation and affordability. Institute for Clinical and Economic Review. https://icer.org/wp-content/uploads/2022/04/ICER-White-Paper_The-Next-Generation-of-Rare-Disease-Drug-Policy_040722.pdf.
Pratt, B. M., L. Hixson, and N. A. Jones. Measuring race and ethnicity across decades, 1790–2010. 2010. Population Division, U.S. Census Bureau. https://www.census.gov/data-tools/demo/race/MREAD_1790_2010.html.
Priddy, F. H., A. C. Cheng, L. F. Salazar, and P. M. Frew. 2006. Racial and ethnic differences in knowledge and willingness to participate in HIV vaccine trials in an urban population in the southeastern US. *International Journal of STD and AIDS* 17(2):99–102.
Protheroe, J., D. Nutbeam, and G. Rowlands. 2009. Health literacy: A necessity for increasing participation in health care. *British Journal of General Practice* 59(567):721–723.
Quinn, S. C. 2004. Ethics in public health research. *American Journal of Public Health* 94(6):918–922. https://doi.org/10.2105/ajph.94.6.918.
Quinn, S. C., J. Butler, C. S. Fryer, M. A. Garza, K. H. Kim, C. Ryan, and S. B. Thomas. 2012. Attributes of researchers and their strategies to recruit minority populations: Results of a national survey. *Contemporary Clinical Trials* 33(6):1231–1237.
Quinn, S. C., A. Jamison, J. An, V. S. Freimuth, G. R. Hancock, and D. Musa. 2018. Breaking down the monolith: Understanding flu vaccine uptake among African Americans. *SSM – Population Health* 4:25–36.
Quiñones, A. R., S. L. Mitchell, J. D. Jackson, M. P. Aranda, P. Dilworth-Anderson, E. P. McCarthy, and L. Hinton. 2020. Achieving health equity in embedded pragmatic trials for people living with dementia and their family caregivers. *Journal of the American Geriatric Society* 68(Suppl. 2):S8–S13.
Ramamoorthy, A., M. A. Pacanowski, J. Bull, and L. Zhang. 2015. Racial/ethnic differences in drug disposition and response: Review of recently approved drugs. *Clinical Pharmacology & Therapeutics* 97(3):263–273.
Ramsey, P. 1976. The enforcement of morals: Nontherapeutic research on children. *Hastings Center Report* 6(4):21–30.
Reihl, S. J., N. Patil, R. A. Morshed, M. Mehari, A. Aabedi, U. N. Chukwueke, A. B. Porter, V. Fontil, G. Cioffi, K. Waite, C. Kruchko, Q. Ostrom, J. Barnholtz-Sloan, and S. L. Hervey-Jumper. 2021. A population study of clinical trial accrual for women and minorities in neuro-oncology following the NIH Revitalization Act. *medRxiv*. https://www.medrxiv.org/content/10.1101/2021.05.28.21258034v1.
Reza, N., A. S. Tahhan, N. Mahmud, E. M. DeFilippis, A. Alrohaibani, M. Vaduganathan, S. J. Greene, A. H. Ho, G. C. Fonarow, J. Butler, C. O'Connor, M. Fiuzat, O. Vardeny, I. L. Piña, J. Lindenfeld, and M. Jessup. 2020. Representation of women authors in international heart failure guidelines and contemporary clinical trials. *Circulation: Heart Failure* 13(8):e006605.
Rhea, B. W., E. G. Lindo, L. D. Weeks, and M. R. McLemore. 2020. On racism: A new standard for publishing on racial health inequities. *Health Affairs*, Health Affairs Blog, posted July 2, 2020. https://www.healthaffairs.org/do/10.1377/forefront.20200630.939347.
Riley, W., M. Riddle, and M. Lauer. 2018. NIH policies on experimental studies with humans. *Nature Human Behavior* 2(2):103–106.
Rivara, F. P., S. M. Bradley, D. V. Catenacci, A. N. Desai, I. Ganguli, S. J. P. A. Haneuse, S. K. Inouye, E. A. Jacobs, K. Kan, H. S. Kim, A. M. Morris, O. Ogedegbe, E. N. Perencevich, R. H. Perlis, E. Powell, G. D. Rubenfeld, L. N. Shulman, N. S. Trueger, and S. D. Fihn. 2021. Structural racism and *JAMA Network Open*. *JAMA Network Open* 4(6):e2120269–e2120269.

Rosende-Roca, M., C. Abdelnour, E. Esteban, J. Tartari, E. Alarcon, J. Martínez-Atienza, A. González-Pérez, M. Sáez, A. Lafuente, M. Buendía, A. Pancho, N. Aguilera, M. Ibarria, S. Diego, S. Jofresa, I. Hernández, R. López, M. Gurruchaga, L. Tárraga, S. Valero, A. Ruiz, M. Marquié, and M. Boada. 2021. The role of sex and gender in the selection of Alzheimer patients for clinical trial pre-screening. *Alzheimer's Research & Therapy* 13(1):95.

RSC (Royal Society for Chemistry). n.d. Talent. Joint commitment for action on inclusion and diversity in publishing. https://www.rsc.org/new-perspectives/talent/joint-commitment-for-action-inclusion-and-diversity-in-publishing/.

Rubin, R. 2021. Pregnant people's paradox—excluded from vaccine trials despite having a higher risk of covid-19 complications. *JAMA* 325(11):1027–1028.

Rucker-Whitaker, C., K. J. Flynn, G. Kravitz, C. Eaton, J. E. Calvin, and L. H. Powell. 2006. Understanding African-American participation in a behavioral intervention: Results from focus groups. *Contemporary Clinical Trials* 27(3):274–286.

Sadler, G. R., J. Gonzalez, M. Mumman, L. Cullen, S. F. Lahousse, V. Malcarne, V. Conde, and N. Riley. 2010. Adapting a program to inform African American and Hispanic American women about cancer clinical trials. *Journal of Cancer Education* 25(2):142–145.

SAMHSA (Substance Abuse and Mental Health Services Administration). 2012. *Top Health Issues for LGBT Populations: Information & Resource Kit.* Washington, DC: U.S. Department of Health and Human Services.

Sanderson, S. C., M. A. Diefenbach, R. Zinberg, C. R. Horowitz, M. Smirnoff, M. Zweig, S. Streicher, E. W. Jabs, and L. D. Richardson. 2013. Willingness to participate in genomics research and desire for personal results among underrepresented minority patients: A structured interview study. *Journal of Community Genetics* 4(4):469–482.

Sauceda, J. A., K. Dubé, B. Brown, A. E. Pérez, C. E. Rivas, D. Evans, and C. B. Fisher. 2021. Framing a consent form to improve consent understanding and determine how this affects willingness to participate in HIV cure research: An experimental survey study. *Journal of Empirical Research on Human Research Ethics* 16(1–2):78–87.

Sausa, L., J. Sevelius, J. Keatley, J. R. Iñiguez, and M. Reyes. 2009. *Policy Recommendations for Inclusive Data Collection of Trans People in HIV Prevention, Care & Services.* San Francisco: Center of Excellence for Transgender HIV Prevention, University of California, San Francisco.

Scaffidi, J., B. W. Mol, and J. A. Keelan. 2017. The pregnant women as a drug orphan: A global survey of registered clinical trials of pharmacological interventions in pregnancy. *BJOG* 124(1):132–140.

Scharff, D. P., K. J. Mathews, P. Jackson, J. Hoffsuemmer, E. Martin, and D. Edwards. 2010. More than Tuskegee: Understanding mistrust about research participation. *Journal of Health Care for the Poor and Underserved* 21(3):879–897.

Scott, P. E., E. F. Unger, M. R. Jenkins, M. R. Southworth, T.-Y. McDowell, R. J. Geller, M. Elahi, R. J. Temple, and J. Woodcock. 2018. Participation of women in clinical trials supporting FDA approval of cardiovascular drugs. *Journal of the American College of Cardiology* 71(18):1960–1969.

Sedrak, M. S., R. A. Freedman, H. J. Cohen, H. B. Muss, A. Jatoi, H. D. Klepin, T. M. Wildes, J. G. Le-Rademacher, G. G. Kimmick, W. P. Tew, K. George, S. Padam, J. Liu, A. R. Wong, A. Lynch, B. Djulbegovic, S. G. Mohile, and W. Dale; Cancer and Aging Research Group. 2021. Older adult participation in cancer clinical trials: A systematic review of barriers and interventions. *CA: A Cancer Journal for Clinicians* 71(1):78–92.

Shepherd, A., D. Hewick, T. Moreland, and I. Stevenson. 1977. Age as a determinant of sensitivity to warfarin. *British Journal of Clinical Pharmacology* 4(3):315–320.

Shields, K. E., and A. D. Lyerly. 2013. Exclusion of pregnant women from industry-sponsored clinical trials. *Obstetrics & Gynecology* 122(5):1077–1081.

Shirkey, H. 1968. Therapeutic orphans. *The Journal of Pediatrics* 72(1):119–120.

Sirugo, G., S. M. Williams, and S. A. Tishkoff. 2019. The missing diversity in human genetic studies. *Cell* 177(1):26–31.

Siskind, R. L., M. Andrasik, S. T. Karuna, G. B. Broder, C. Collins, A. Liu, J. P. Lucas, G. W. Harper, and P. O. Renzullo. 2016. Engaging transgender people in NIH-funded HIV/AIDS clinical trials research. *Journal of Acquired Immune Deficiency Syndromes (1999)* 72(Suppl. 3):S243.

Siskind, R. L., M. Andrasik, S. T. Karuna, G. B. Broder, C. Collins, A. Liu, J. P. Lucas, G. W. Harper, and P. O. Renzullo. 2016. Engaging transgender people in nih-funded hiv/aids clinical trials research. *Journal of Acquired Immune Deficiency Syndromes (1999)* 72(Suppl. 3):S243.

Skirrow, H., S. Barnett, S. Bell, S. L. Riaposova, S. Mounier-Jack, B. Kampmann, and B. Holder. 2022. Women's views on accepting COVID-19 vaccination during and after pregnancy, and for their babies: a multi-methods study in the UK. *BMC Pregnancy Childbirth* 22:33. https://doi.org/10.1186/s12884-021-04321-3.

Slomka, J., E. A. Ratliff, S. A. McCurdy, S. Timpson, and M. L. Williams. 2008. Decisions to participate in research: Views of underserved minority drug users with or at risk for HIV. *AIDS Care* 20(10):1224–1232.

Smirnoff, M., I. Wilets, D. F. Ragin, R. Adams, J. Holohan, R. Rhodes, G. Winkel, E. M. Ricci, C. Clesca, and L. D. Richardson. 2018. A paradigm for understanding trust and mistrust in medical research: The community voices study. *AJOB Empirical Bioethics* 9(1):39–47.

Smith, S. K., W. Selig, M. Harker, J. N. Roberts, S. Hesterlee, D. Leventhal, R. Klein, B. Patrick-Lake, and A. P. Abernethy. 2015. Patient engagement practices in clinical research among patient groups, industry, and academia in the United States: A survey. *PLOS One* 10(10):e0140232.

Soejima K., H. Sato, and A. Hisaka. 2022. Age-related change in hepatic clearance inferred from multiple population pharmacokinetic studies: Comparison with renal clearance and their associations with organ weight and blood flow. *Clinical Pharmacokinetics* 61(2):295–305. doi: 10.1007/s40262-021-01069-z. Epub 2021 Sep 13. PMID: 34514537.

Spong, C. Y., and D. W. Bianchi. 2018. Improving public health requires inclusion of underrepresented populations in research. *JAMA* 319(4):337–338.

Sprague, D., J. Russo, D. L. Lavallie, and D. Buchwald. 2013. Barriers to cancer clinical trial participation among American Indian and Alaska Native tribal college students. *Journal of Rural Health* 29(1):55–60.

Stadnick N. A., K. L. Cain, W. Oswald, P. Watson, M. Ibarra, R. Lagoc, L. O. Ayers, L. Salgin, S. L. Broyles, L. C. Laurent, K. Pezzoli, and B. Rabin. Co-creating a Theory of Change to advance COVID-19 testing and vaccine uptake in underserved communities. Health Serv Res. 2022 Mar 4. doi:10.1111/1475-6773.13910.

Stefanoudis, P. V., W. Y. Licuanan, T. H. Morrison, S. Talma, J. Veitayaki, and L. C. Woodall. 2021. Turning the tide of parachute science. *Current Biology* 31(4):R184–R185.

Stephenson, J. 2020. FDA offers guidance for boosting diversity in clinical trials. *JAMA Health Forum* 1(11):e201434.

Stevens, K. R., K. S. Masters, P. I. Imoukhuede, K. A. Haynes, L. A. Setton, E. Cosgriff-Hernandez, M. A. Lediju Bell, P. Rangamani, S. E. Sakiyama-Elbert, S. D. Finley, R. K. Willits, A. N. Koppes, N. C. Chesler, K. L. Christman, J. B. Allen, J. Y. Wong, H. El-Samad, T. A. Desai, and O. Eniola-Adefeso. 2021. Fund Black scientists. *Cell* 184(3):561–565.

Strauss, R. P., S. Sengupta, S. C. Quinn, J. Goeppinger, C. Spaulding, S. M. Kegeles, and G. Millett. 2001. The role of community advisory boards: Involving communities in the informed consent process. *American Journal of Public Health*. 91(12):1938–1943. doi:10.2105/ajph.91.12.1938.

Strong, B., V. Howard, and M. Reeves. 2020. Sex differences in enrollment among contemporary acute stroke trials: Differences by trial type. Paper presented at ESO-WSO 2020 Virtual Conference, November 7–9, 2020. ESO-WSO Joint Meeting Abstracts. *International Journal of Stroke*. https://journals.sagepub.com/doi/full/10.1177/1747493020963387.

Sullivan, J. 2004. Subject recruitment and retention: Barriers to success. *Applied Clinical Trials* 13(4).

Szabo, L. 2022. Why pregnant people were left behind while vaccines moved at 'warp speed' to help the masses. *Kaiser Health News*. https://khn.org/news/article/why-pregnant-people-were-left-behind-while-vaccines-moved-at-warp-speed-to-help-the-masses/.

Tahhan A. S., M. Vaduganathan, S. J. Greene, G. C. Fonarow, M. Fiuzat, M. Jessup, J. Lindenfeld, C. M. O'Connor, and J. Butler. 2018. Enrollment of older patients, women, and racial and ethnic minorities in contemporary heart failure clinical trials: A systematic review. *JAMA Cardiology* 3(10):1011–1019. doi: 10.1001/jamacardio.2018.2559. PMID: 30140928.

Tahhan A. S., M. Vaduganathan, S. J. Greene, A. Alrohaibani, M. Raad, M. Gafeer, R. Mehran, G. C. Fonarow, P. S. Douglas, D. L. Bhatt, and J. Butler. 2020. Enrollment of older patients, women, and racial/ethnic minority groups in contemporary acute coronary syndrome clinical trials: A systematic review. *JAMA Cardiology* 5(6):714–722. doi:10.1001/jamacardio.2020.0359. PMID: 32211813.

Takvorian, S. U., C. E. Guerra, and W. L. Schpero. 2021. A hidden opportunity — Medicaid's role in supporting equitable access to clinical trials. *New England Journal of Medicine* 384(21): 1975–1978.

Tamargo, J., G. Rosano, T. Walther, J. Duarte, A. Niessner, J. C. Kaski, C. Ceconi, H. Drexel, K. Kjeldsen, and G. Savarese. 2017. Gender differences in the effects of cardiovascular drugs. *European Heart Journal–Cardiovascular Pharmacotherapy* 3(3):163–182.

Tauer, C. A. 2002. Central ethical dilemmas in research involving children. *Accountability in Research: Policies and Quality Assurance* 9(3–4):127–142.

Tay, T., J. Pham, and M. Hew. 2020. Addressing the impact of ethnicity on asthma care. *Current Opinion in Allergy and Clinical Immunology* 20(3):274–281.

Taylor, H. A. 2008. Implementation of NIH inclusion guidelines: Survey of NIH study section members. *Clinical Trials* 5(2):140–146.

TCFHA (The Center for Health Affairs). 2012. The emerging field of patient navigation: A golden opportunity to improve healthcare. Issue Brief, December 20, 2012. https://www.neohospitals.org/thecenterforhealthaffairs/mediacenter/newsreleases/2012/December/12-12_Patient-Navigation-Publication.

Thetford, K., T. W. Gillespie, Y. I. Kim, B. Hansen, and I. C. Scarinci. 2021. Willingness of Latinx and African Americans to participate in nontherapeutic trials: It depends on who runs the research. *Ethnicity and Disease* 31(2):263–272.

The White House. 2000. Office of the Press Secretary. Medicare will reimburse for all routine patient care costs for those in clinical trials. Press Announcement, June 7, 2000. https://clintonwhitehouse5.archives.gov/WH/New/html/20000607.html.

Torgersen J., S. L. Bellamy, B. Ratshaa, X. Han, M. Mosepele, A. F. Zuppa, M. Vujkovic, A. P. Steenhoff, G. P. Bisson, and R. Gross. 2019. Impact of efavirenz metabolism on loss to care in older HIV+ africans. *European Journal of Drug Metabolism and Pharmacokinetics* 44(2):179–187.

Trant, A. A., L. Walz, W. Allen, J. DeJesus, C. Hatzis, and A. Silber. 2020. Increasing accrual of minority patients in breast cancer clinical trials. *Breast Cancer Research and Treatment* 184(2):499–505.

Trantham, L. C., W. R. Carpenter, L. D. DiMartino, B. White, M. Green, R. Teal, G. Corbie-Smith, and P. A. Godley. 2015. Perceptions of cancer clinical research among African American men in North Carolina. *Journal of the National Medical Association* 107(1):33–41.

Tu, S. P., H. Chen, A. Chen, J. Lim, S. May, and C. Drescher. 2005. Clinical trials: Understanding and perceptions of female Chinese-American cancer patients. *Cancer* 104(12 Suppl.):2999–3005.

UCSF Accelerate. 2022. How can I best work with community members to enhance my research and its impact on the community?, University of California, San Francisco. https://accelerate.ucsf.edu/files/CE/communityFAQ.pdf.

UNC (University of North Carolina at Chapel Hill). 2022. Atherosclerosis risk in communities study description. https://sites.cscc.unc.edu/aric/desc.

Unger, J. M., J. R. Gralow, K. S. Albain, S. D. Ramsey, and D. L. Hershman. 2016. Patient income level and cancer clinical trial participation. *JAMA Oncology* 2(1):137.

USPSTF (U.S. Preventive Services Task Force). 2021. Screening for colorectal cancer: U.S. Preventive Services Task Force recommendation statement. *JAMA* 325(19):1965–1977.

van der Kolk, B. A. 2014. *The Body Keeps the Score: Brain, Mind, and Body in the Healing of Trauma*. New York: Viking.

Van Nuys, K. E., Z. Xie, B. Tysinger, M. A. Hlatky, and D. P. Goldman. 2018. Innovation in heart failure treatment: Life expectancy, disability, and health disparities. *JACC: Heart Failure* 6(5):401–409.

Van Spall, H. G. C. 2021. Exclusion of pregnant and lactating women from COVID-19 vaccine trials: A missed opportunity. *European Heart Journal* 42(28):2724–2726.

Vargesson, N. 2015. Thalidomide-induced teratogenesis: History and mechanisms. *Birth Defects Research Part C: Embryo Today: Reviews* 105(2):140–156.

Vigil, D., N. Sinaii, and B. Karp. 2021. American Indian and Alaska Native enrollment in clinical studies in the National Institutes of Health's Intramural Research Program. *Ethics & Human Research* 43(3):2–9.

Vitale, C., M. Fini, I. Spoletini, M. Lainscak, P. Seferovic, and G. Rosano. 2017. Under-representation of elderly and women in clinical trials. *International Journal of Cardiology* 232:216–221.

Vitale, C., G. Rosano, and M. Fini. 2016. Are elderly and women under-represented in cardiovascular clinical trials? Implication for treatment. *Wiener klinische Wochenschrift* 128(Suppl. 7).

Voytek, C. D., K. T. Jones, and D. S. Metzger. 2011. Selectively willing and conditionally able: HIV vaccine trial participation among women at "high risk" of HIV infection. *Vaccine* 29(36):6130–6135.

Waheed, W., A. Hughes-Morley, A. Woodham, G. Allen, and P. Bower. 2015. Overcoming barriers to recruiting ethnic minorities to mental health research: A typology of recruitment strategies. *BMC Psychiatry* 15(1).

Wallace, J., K. Jiang, P. Goldsmith-Pinkham, and Z. Song. 2021. Changes in racial and ethnic disparities in access to care and health among US adults at age 65 years. *JAMA Internal Medicine* 181(9):1207–1215.

Wallerstein, N. B., and B. Duran. 2006. Using community-based participatory research to address health disparities. *Health Promotion Practice* 7(3):312–323.

Wallerstein, N., B. Duran, J. Oetzel, and M. Minkler, eds. 2018. *Community-based Participatory Research for Health: Advancing Social and Health Equity*. 3rd ed. San Francisco, CA: Jossey-Bass.

Wang, R., S. W. Lagakos, J. H. Ware, D. J. Hunter, and J. M. Drazen. 2007. Statistics in medicine—reporting of subgroup analyses in clinical trials. *New England Journal of Medicine* 357(21):2189–2194.

Ward, E., A. Jemal, V. Cokkinides, G. K. Singh, C. Cardinez, A. Ghafoor, and M. Thun. 2004. Cancer disparities by race/ethnicity and socioeconomic status. *CA: A Cancer Journal for Clinicians* 54(2):78–93.

Warden, B. A., S. Fazio, and M. D. Shapiro. 2020. The PCSK9 revolution: Current status, controversies, and future directions. *Trends in Cardiovascular Medicine* 30(3):179–185.

Webb, D. A., J. C. Coyne, R. L. Goldenberg, V. K. Hogan, I. T. Elo, J. R. Bloch, L. Mathew, I. M. Bennett, E. F. Dennis, and J. F. Culhane. 2010. Recruitment and retention of women in a large randomized control trial to reduce repeat preterm births: The Philadelphia Collaborative Preterm Prevention Project. *BMC Medical Research Methodology* 10.

Webb, F. J., J. Khubchandani, C. W. Striley, and L. B. Cottler. 2019. Black–white differences in willingness to participate and perceptions about health research: Results from the population-based healthstreet study. *Journal of Immigrant and Minority Health* 21(2):299–305.

Wendler, D., R. Kington, J. Madans, G. Van Wye, H. Christ-Schmidt, L. A. Pratt, O. W. Brawley, C. P. Gross, and E. Emanuel. 2006. Are racial and ethnic minorities less willing to participate in health research? *PLOS Medicine* 3(2):201–210.

Wennberg, D. E., F. L. Lucas, J. D. Birkmeyer, C. E. Bredenberg, and E. S. Fisher. 1998. Variation in carotid endarterectomy mortality in the Medicare population: Trial hospitals, volume, and patient characteristics. *JAMA* 279(16):1278–1281. doi:10.1001/jama.279.16.1278.

Wesp, L. M., V. Scheer, A. Ruiz, K. Walker, J. Weitzel, L. Shaw, P. M. Kako, and L. Mkandawire-Valhmu. 2018. An emancipatory approach to cultural competency: The application of critical race, postcolonial, and intersectionality theories. *Advances in Nursing Science* 41(4):316–326.

Westergaard, R. P., M. C. Beach, S. Saha, and E. A. Jacobs. 2014. Racial/ethnic differences in trust in health care: HIV conspiracy beliefs and vaccine research participation. *Journal of General Internal Medicine* 29(1):140–146.

Whitelaw, S., K. Sullivan, Y. Eliya, M. Alruwayeh, L. Thabane, C. W. Yancy, R. Mehran, M. A. Mamas, and H. G. C. Van Spall. 2021. Trial characteristics associated with under-enrolment of females in randomized controlled trials of heart failure with reduced ejection fraction: A systematic review. *European Journal of Heart Failure* 23(1):15–24.

Wilcox, A., K. Natarajan, and C. Weng. 2009. Using personal health records for automated clinical trials recruitment: The ePaIRing Model. *Summit on Translational Bioinformatics* 2009:136–140.

Wilkins, C. H. 2018. Effective engagement requires trust and being trustworthy. *Medical Care* 56(10 Suppl. 1):S6–S8.

Wilkins, C. H., and P. M. Alberti. 2019. Shifting academic health centers from a culture of community service to community engagement and integration. *Academic Medicine* 94(6):763–767.

Winkfield, K. M., J. K. Phillips, S. Joffe, M. T. Halpern, D. S. Wollins, and B. Moy. 2018. Addressing financial barriers to patient participation in clinical trials: ASCO policy statement. *Journal of Clinical Oncology* 6(33):3331–3339. doi:10.1200/JCO.18.01132.

Winter, S. S., J. M. Page-Reeves, K. A. Page, E. Haozous, A. Solares, C. Nicole Cordova, and R. S. Larson. 2018. Inclusion of special populations in clinical research: Important considerations and guidelines. *Journal of Clinical and Translational Research* 4(1):56–69.

Wissing, M.D., Kluetz, P.G., Ning, Y.-M., Bull, J., Merenda, C., Murgo, A.J. and Pazdur, R. 2014. Under-representation of racial minorities in prostate cancer studies submitted to the US Food and Drug Administration to support potential marketing approval, 1993-2013. *Cancer* 120: 3025–3032.

WMA (World Medical Association). 2008. Declaration of Helsinki. https://www.wma.net/wp-content/uploads/2018/07/DoH-Oct2008.pdf.

Woitowich, N. C., and T. K. Woodruff. 2019. Opinion: Research community needs to better appreciate the value of sex-based research. *Proceedings of the National Academy of Sciences* 116(15):7154–7156.

Women's health. 1985. Report of the Public Health Service Task Force on Women's Health Issues. *Public Health Reports* 100(1):73–106.

Wood, S. 2021. Improving the representation of women and underrepresented minorities in clinical trials and research. In *NASEM Report Interviews*, L. Bothwell, ed.

Woodall, A., C. Morgan, C. Sloan, and L. Howard. 2010. Barriers to participation in mental health research: Are there specific gender, ethnicity and age related barriers? *BMC Psychiatry* 10(1):103.

Yancey, A. K., A. N. Ortega, and S. K. Kumanyika. 2006. Effective recruitment and retention of minority research participants. *Annual Review of Public Health* 27(1):1–28.

Yearby, R. 2014. When is a change going to come. Separate and unequal treatment in health care fifty years after Title VI of the Civil Rights Act of 1964. *SMU Law Review* 67(2):287–337.

Yin, W. 2008. Market incentives and pharmaceutical innovation. *Journal of Health Economics* 27(4). https://doi.org/10.1016/j.jhealeco.2008.01.002.

Zissimopoulos, J. M., B. C. Tysinger, P. A. St. Clair, and E. M. Crimmins. 2018. The impact of changes in population health and mortality on future prevalence of Alzheimer's disease and other dementias in the united atates. *The Journals of Gerontology: Series B* 73(Suppl. 1):S38–S47.

Zulman, D. M., J. B. Sussman, X. Chen, C. T. Cigolle, C. S. Blaum, and R. A. Hayward. 2011. Examining the evidence: A systematic review of the inclusion and analysis of older adults in randomized controlled trials. *Journal of General Internal Medicine* 26(7):783–790.

Zunzunegui, M., B. Alvarado, R. Guerra, J. Gómez, A. Ylli, J. Guralnik, and I. R. Group. 2015. The mobility gap between older men and women: The embodiment of gender. *Archives of Gerontology and Geriatrics* 61(2):140–148.

Appendix A

Quantifying the Potential Health and Economic Impacts of Increased Trial Diversity

Bryan Tysinger[1]

INTRODUCTION

Chronic illness decreases quantity of life, quality of life, and years spent in the labor force. Less appreciated is the potential for differential impact of disease for different race/ethnicity-gender groups. In other words, while chronic illness affects outcomes for all groups, some groups might experience a larger impact. The goal in this analysis is to quantify the differential impact of chronic illness for groups that have historically been underrepresented in clinical trials, as clinical trials are a potential way to identify approaches to reduce these disparities. We examine three key outcomes: quantity of life (measured by life expectancy), quality of life (measured by disability-free life), and working life (measured by years in the labor force). The thought experiment considers a hypothetical world where the differential impact is eliminated, that is, that all groups share the same impact of chronic illness.

To do this, we utilize a dynamic microsimulation model, the Future Elderly Model (FEM), to project a baseline scenario for groups of interest for each of three chronic conditions. We then consider a counterfactual scenario in which disparities in disease impact on mortality, disability, and workforce participation are eliminated.

Future Elderly Model

The Future Elderly Model is a dynamic microsimulation of health risk factors, chronic illnesses, disability, and health-related economic outcomes for the U.S. population over the age of 50. It simulates the aging process for individuals, including projecting risk factors like smoking and BMI (body mass

---

[1] Available at btysinge@usc.edu.

index), chronic conditions like diabetes and heart disease, functional limitations in Activities of Daily Living (ADLs) and Instrumental Activities of Daily Living (IADLs), and economic outcomes such as workforce participation and medical spending. FEM relies on statistical models based on real individuals who participate in a nationally representative panel survey.

The FEM has been used in support of a broad set of research. A previous National Academies of Sciences, Engineering, and Medicine report relied on FEM analyses to quantify the impact of growing disparities in life expectancy on federal programs (NASEM, 2015). Early work with the microsimulation explored trends in health, the value of prevention, and the resulting fiscal consequences (Goldman et al., 2005, 2009, 2010; Lakdawalla et al., 2005). More recent work has targeted disparities and innovation in particular diseases such as congestive heart failure and Alzheimer's disease (Van Nuys et al., 2018; Zissimopoulos et al., 2018). Crucially, projections from FEM have been extensively validated (Leaf et al., 2020).

Data

This analysis utilizes the Health and Retirement Study (HRS), a nationally representative panel study of Americans over the age of 50. The HRS is sponsored by the National Institute on Aging (grant number NIA U01AG009740) and is conducted by the University of Michigan (RAND HRS, 2021a; RAND HRS, 2021b).

Groups of Interest

We identified six groups of interest in the HRS with sufficient sample size to support this analysis. Throughout, non-Hispanic white males serve as the reference group due to their historical inclusion and representation in clinical trials. Non-Hispanic Black males, Hispanic males, non-Hispanic white females, non-Hispanic Black females, and Hispanic females all potentially benefit from narrowing the differential impact of disease on the outcomes of interest.

Diseases of Interest

We considered three types of chronic conditions that come from self-reported data in the HRS: diabetes, heart diseases, and hypertension. A person is identified as having diabetes based on the question, "Has a doctor ever told you that you have diabetes or high blood sugar?" Heart diseases includes a broad set of conditions that affect the heart. This is based on the question, "Has a doctor ever told you that you have had a heart attack, coronary heart disease, angina, congestive heart failure, or other heart problems?" Hypertension is based on the question, "Has a doctor ever told you that you have high blood pressure or hypertension?"

Due to the wording of these questions, we consider them absorbing states. That is, once a person indicates they were diagnosed with a condition, then they have the condition for the remainder of their life.

Outcomes of Interest

We focused on three key outcomes of interest: mortality, disability, and working for pay. Mortality in the HRS is measured by proxy response. Since the HRS is collected every 2 years, mortality is modeled as 2-year mortality incidence. Disability is a composite measure based on limitations in ADLs, IADLs, or living in a nursing home. If the respondent reports any ADLs, any IADLs, or living in a nursing home, they are considered a person with a disability. Working for pay is derived from self-reported status of working for pay and labor force participation.

Estimation

Transition models are the statistical models that drive the microsimulation. The transition models for disease incidence in the FEM rely on a first-order Markov structure. As such, any time-varying predictors enter as "lagged" variables from the previous wave of the survey. Time-varying predictors include things like BMI, smoking status, and other chronic conditions.

Diabetes incidence is modeled as a function of gender, race, age, BMI, and smoking. Hypertension incidence has a similar structure, but also controls for diabetes. Similarly, heart disease incidence controls for these variables, but also controls for diabetes and hypertension. Risk factors like smoking and BMI are also transitioned within the simulation.

The three key outcomes of interest—mortality, disability, and work—are estimated with a "reduced form" approach. For each disease of interest, transition models for these outcomes are functions of group, group-specific age profiles, the disease, and an underrepresented group indicator variable interacted with the disease. This last term is the key parameter of interest. If this parameter were zero, it would indicate no disparity between the reference group (non-Hispanic white males) and the underrepresented groups.

Transition models are estimated using the HRS respondents' data from 1998 to 2018. Sample characteristics for the 2018 sample are shown below (see Table A-1).

The parameter estimates and marginal effects for the key transition models are shown in Tables A-11, A-12, and A-13. Adjusted relative risks for the key parameters of interest (the underrepresented group and disease interaction term) are shown in A-2. The reference group, non-Hispanic white males, will always have values of 1.0. Relative to white males, being in an underrepresented group and having diabetes is associated with an increase in mortality of 10 to 11 percent, an increase in disability of 10 to 12 percent, and a decrease in workforce partici-

TABLE A-1 1998–2018 Health and Retirement Study Sample Characteristics

	Mean	SD
Age	69.0	10.9
Non-Hispanic white males	30%	0.46
Non-Hispanic Black males	6%	0.24
Hispanic males	5%	0.21
Non-Hispanic white females	41%	0.49
Non-Hispanic Black females	11%	0.31
Hispanic females	7%	0.25
BMI	28.0	6.0
Ever smoke	57%	0.50
Current smoker	13%	0.34
Ever had diabetes	22%	0.41
Ever had heart disease	25%	0.43
Ever had hypertension	57%	0.49
Any disability	22%	0.41
Working for pay	35%	0.48
Died	6%	0.24
N = 191,036		

pation of 9 to 12 percent. Heart disease is associated with a mortality increase of 14 to 15 percent, an increase in disability of 19 to 23 percent, and a decrease in workforce participation of 11 to 14 percent. Hypertension is associated with an increase in mortality of 10 to 11 percent, an increase in disability of 14 to 17 percent, and a decrease in workforce participation of 4 to 5 percent.

Simulation

Table A-3 shows the baseline characteristics for the 2016 cohorts of 51–52-year-olds at the start of the simulation. Initial prevalence of disease varies across groups, with the highest rates of diabetes among non-Hispanic Black males, Hispanic males, and Hispanic females. Heart disease at baseline is highest among non-Hispanic white females and non-Hispanic Black males. Hypertension rates are highest for non-Hispanic Black males and females. Rates of disability are higher for females, and workforce participation is higher among males.

TABLE A-2 Adjusted Relative Risks for Key Parameters of Interest

	Diabetes				Heart Disease				Hypertension		
	Mortality	Disability	Work		Mortality	Disability	Work		Mortality	Disability	Work
White males	1.00	1.00	1.00		1.00	1.00	1.00		1.00	1.00	1.00
Black males	1.10 [1.02, 1.18]	1.12 [1.07, 1.16]	0.89 [0.85, 0.92]		1.14 [1.07, 1.22]	1.23 [1.18, 1.27]	0.86 [0.83, 0.90]		1.10 [1.02, 1.19]	1.17 [1.13, 1.22]	0.95 [0.93, 0.98]
Hispanic males	1.11 [1.02, 1.20]	1.12 [1.07, 1.16]	0.91 [0.88, 0.94]		1.15 [1.07, 1.23]	1.22 [1.18, 1.27]	0.89 [0.86, 0.92]		1.11 [1.03, 1.20]	1.17 [1.12, 1.21]	0.96 [0.94, 0.98]
White females	1.10 [1.02, 1.19]	1.11 [1.07, 1.16]	0.89 [0.85, 0.92]		1.14 [1.07, 1.21]	1.21 [1.17, 1.26]	0.86 [0.82, 0.90]		1.10 [1.02, 1.18]	1.16 [1.12, 1.20]	0.95 [0.92, 0.98]
Black females	1.11 [1.02, 1.20]	1.10 [1.06, 1.14]	0.88 [0.85, 0.92]		1.15 [1.07, 1.23]	1.19 [1.15, 1.22]	0.86 [0.83, 0.90]		1.11 [1.03, 1.20]	1.15 [1.11, 1.19]	0.95 [0.93, 0.98]
Hispanic females	1.11 [1.02, 1.21]	1.10 [1.06, 1.14]	0.88 [0.85, 0.92]		1.15 [1.07, 1.23]	1.18 [1.15, 1.22]	0.86 [0.82, 0.90]		1.11 [1.03, 1.20]	1.14 [1.11, 1.18]	0.95 [0.92, 0.98]

TABLE A-3 Baseline Characteristics at Simulation Start

	Non-Hispanic White Males	Non-Hispanic White Females	Non-Hispanic Black Males	Non-Hispanic Black Females	Hispanic Males	Hispanic Females
Weighted N	2,879,983	2,920,961	509,836	576,820	648,817	633,641
Age	52	52	52	52	52	52
BMI	29.3	30.6	30.7	33.3	29.9	30.7
Current smoker	25%	16%	23%	19%	21%	24%
Diabetes	14%	11%	23%	13%	26%	29%
Heart disease	8%	15%	10%	6%	3%	7%
Hypertension	39%	33%	57%	55%	38%	38%
Any disability	18%	20%	15%	17%	11%	20%
Working for pay	81%	79%	72%	66%	89%	65%

APPENDIX A

PROJECTIONS

Diabetes

In the baseline scenario, average life expectancy for those who develop diabetes prior to death ranges from 27.2 years (non-Hispanic Black males) to 34.0 years (Hispanic females). Eliminating the underrepresented group diabetes effect increases life expectancy by 0.8 to 0.9 years in the counterfactual scenario. Similarly, disability-free life increases by 1.0 to 1.2 years, and workforce participation increases by 0.4 to 0.6 years (see Table A-4).

Heart Disease

Baseline and counterfactual projections for the heart disease scenarios are shown in Table A-5. Life expectancy increases between 0.9 and 1.1 years for the underrepresented groups. Disability-free life years increase from 1.4 to 1.6 years. Years working increase from 0.2 to 0.4 years.

Hypertension

As seen in Table A-6 in the hypertension scenarios, life expectancy increases 0.9 to 1.1 years when the underrepresentation gap is eliminated. Disability-free life years increase from 1.4 to 1.7 years. Years working increase between 0.3 and 0.4 years.

Valuing the Potential Gains

To value the potential gains in the counterfactual scenarios, we multiplied the number of individuals in the group, their lifetime risk of the disease, the potential change in the outcome of interest, and valued the gain at a commonly used amount. For life years and disability-free life years, we used $150,000 per year. For earnings, we used $50,000 per year. All future benefits are discounted at 3 percent per year.

Lifetime risk for developing these chronic illnesses is high for the 51–52-year-old cohort in the FEM, as seen in Table A-7, Table A-8, and Table A-9. Diabetes risk ranges from 47 percent for non-Hispanic white females to 77 percent for Hispanic females. Heart disease risk ranges from 57 percent for non-Hispanic Black males to 68 percent for non-Hispanic white females. Hypertension risk is high for all groups.

In aggregate, the potential value in narrowing the disparity in chronic disease outcomes is large. For diabetes (see Table A-7), the total impact associated with life expectancy is $128.5 billion. The value is larger for disability-free life expectancy, at $202.5 billion. Additional working years aggregate to $40.6 billion in foregone wages.

TABLE A-4 Life Years, Disability-free Life Years, and Remaining Work Years for Diabetes Scenario

	Baseline	Conterfactual	Delta	Baseline	Conterfactual	Delta	Baseline	Counterfactual	Delta
Hispanic females	34.0 [33.7, 34.3]	34.9 [34.6, 35.2]	0.9 [0.9, 0.9]	21.6 [21.5, 21.7]	22.8 [22.7, 22.9]	1.2 [1.1, 1.3]	7.9 [7.9, 7.9]	8.3 [8.3, 8.3]	0.5 [0.5, 0.5]
Hispanic males	30.2 [30.1, 30.3]	31.1 [31.0, 31.2]	09 [0.8, 1.0]	22.5 [22.4, 22.6]	23.7 [23.6, 23.8]	1.2 [1.1, 1.3]	11.7 [11.6, 11.8]	12.3 [12.2, 12.4]	0.6 [0.6, 0.6]
Non-Hispanic Black females	31.1 [30.9, 31.3]	32.0 [31.8, 32.2]	0.9 [0.8, 1.0]	20.8 [20.6, 21.0]	21.8 [21.6, 22.0]	1.0 [0.9, 1.1]	9.2 [9.2, 9.2]	9.7 [9.7, 9.7]	0.5 [0.5, 0.5]
Non-Hispanic Black males	27.2 [27.1, 27.3]	28.1 [28.0, 28.2]	0.9 [0.8, 1.0]	20.9 [20.8, 21.0]	22.1 [22.0, 22.2]	1.1 [1.0, 1.2]	9.9 [9.8, 10.0]	10.5 [10.4, 10.6]	0.6 [0.6, 0.6]
Non-Hispanic white females	32.9 [32.8, 33.0]	33.7 [33.6, 33.8]	0.8 [0.8, 0.8]	25.4 [25.4, 25.4]	26.4 [26.3, 26.5]	1.0 [1.0, 1.0]	10.4 [10.4, 10.4]	10.8 [10.8, 10.8]	0.4 [0.4, 0.4]
Non-Hispanic white males	30.5 [30.4, 30.6]			27.0 [27.0, 27.0]			13.3 [13.3, 13.3]		

TABLE A-5 Life Years, Disability-free Life Years, and Remaining Work Years for Heart Disease Scenario

	Baseline	Conterfactual	Delta	Baseline	Conterfactual	Delta	Baseline	Counterfactual	Delta
Hispanic females	36.3 [36.3, 36.9]	37.7 [37.3, 38.1]	1.0 [0.9, 1.1]	23.2 [23.1, 23.3]	24.5 [24.4, 24.6]	1.4 [1.4, 1.4]	8.3 [8.3, 8.3]	8.6 [8.6, 8.6]	0.2 [0.2, 0.2]
Hispanic males	33.6 [33.5, 33.7]	34.5 [34.4, 34.6]	0.9 [0.9, 0.9]	25.1 [25.0, 25.2]	26.4 [26.3, 26.5]	1.4 [1.4, 1.4]	12.6 [12.5, 12.7]	12.8 [12.7, 12.9]	0.3 [0.3, 0.3]
Non-Hispanic Black females	34.2 [33.8, 34.6]	35.2 [34.8, 35.6]	1.0 [0.9, 1.1]	22.7 [22.3, 23.1]	24.1 [23.7, 24.5]	1.4 [1.3, 1.5]	9.7 [9.7, 9.7]	10.0 [10.0, 10.0]	0.3 [0.3, 0.3]
Non-Hispanic Black males	30.2 [30.1, 30.3]	31.2 [31.1, 31.3]	1.0 [1.0, 1.0]	23.2 [23.1, 23.3]	24.7 [24.6, 24.8]	1.5 [1.5, 1.5]	10.4 [10.3, 10.5]	10.8 [10.7, 10.9]	0.4 [0.4, 0.4]

	Baseline	Counterfactual	Delta	Baseline	Counterfactual	Delta	Baseline	Counterfactual	Delta
Non-Hispanic white females	35.0 [34.9, 35.1]	36.1 [36.0, 36.2]	1.1 [1.0, 1.2]	27.0 [26.9, 27.1]	28.6 [28.5, 28.7]	1.6 [1.6, 1.6]	10.7 [10.7, 10.7]	11.1 [11.1, 11.1]	0.4 [0.4, 0.4]
Non-Hispanic white males	33.0 [33.0, 33.0]			27.7 [27.7, 27.7]			14.0 [14.0, 14.0]		

TABLE A-6 Life Years, Disability-free Life Years, and Remaining Work Years for Hypertension Scenario

	Baseline	Counterfactual	Delta	Baseline	Counterfactual	Delta	Baseline	Counterfactual	Delta
Hispanic females	35.9 [35.6, 36.2]	36.9 [36.6, 37.2]	1.0 [0.9, 1.1]	23.6 [23.5, 23.7]	25.2 [25.1, 25.3]	1.6 [1.5, 1.7]	8.4 [8.4, 8.4]	8.6 [8.6, 8.6]	0.3 [0.3, 0.3]
Hispanic males	31.6 [31.5, 31.7]	32.6 [32.5, 32.7]	1.0 [0.9, 1.1]	24.3 [24.2, 24.4]	25.9 [25.8, 26.0]	1.6 [1.5, 1.7]	12.3 [12.2, 12.4]	12.6 [12.5, 12.7]	0.3 [0.3, 0.3]
Non-Hispanic Black females	31.9 [31.6, 32.2]	33.0 [32.8, 33.2]	1.1 [1.0, 1.2]	22.2 [21.9, 22.5]	23.9 [23.6, 24.2]	1.7 [1.6, 1.8]	9.6 [9.6, 9.6]	9.9 [9.9, 9.9]	0.4 [0.4, 0.4]
Non-Hispanic Black males	27.9 [27.8, 28.0]	28.9 [28.8, 29.0]	1.0 [0.9, 1.1]	22.3 [22.2, 22.4]	24.0 [23.9, 24.1]	1.6 [1.5, 1.7]	10.3 [10.2, 10.4]	10.7 [10.6, 10.8]	0.4 [0.4, 0.4]
Non-Hispanic white females	34.8 [34.7, 34.9]	35.7 [35.6, 35.8]	0.9 [0.8, 1.0]	27.6 [27.5, 27.7]	29.0 [28.9, 29.1]	1.4 [1.4, 1.4]	11.0 [11.0, 11.0]	11.3 [11.3, 11.3]	0.2 [0.3, 0.3]
Non-Hispanic white males	31.4 [31.3, 31.5]			26.7 [26.7, 26.7]			13.6 [13.6, 13.6]		

TABLE A-7 Aggregate Value of Diabetes Scenario

	N	Lifetime diabetes risk	LE (discounted)	DFLY (discounted)	Work years (discounted)	Aggregate LE	Aggregate DFLY	Aggregate WY
Hispanic females	633,641	77%	0.29 [0.27, 0.30]	0.50 [0.48, 0.53]	0.28 [0.27, 0.29]	$20.9 [$19.7, $22.1]	$36.6 [$34.8, $38.4]	$6.7 [$6.4, $7.0]
Hispanic males	648,817	71%	0.30 [0.27, 0.32]	0.49 [0.46, 0.51]	0.32 [0.31, 0.34]	$20.5 [$19.0, $22.0]	$33.9 [$32.2, $35.6]	$7.5 [$7.1, $7.8]
Non-Hispanic Black females	576,820	63%	0.30 [0.27, 0.32]	0.43 [0.41, 0.46]	0.25 [0.24, 0.26]	$16.2 [$15.1, $17.3]	$23.7 [$22.3, $25.0]	$4.6 [$4.3, $4.8]
Non-Hispanic Black males	509,836	65%	0.32 [0.30, 0.34]	0.48 [0.46, 0.50]	0.33 [0.31, 0.34]	$15.9 [$14.8, $17.0]	$23.9 [$22.8, $24.9]	$5.4 [$5.2, $5.6]
Non-Hispanic white females	2,920,961	47%	0.27 [0.25, 0.28]	0.41 [0.39, 0.43]	0.24 [0.23, 0.25]	$54.9 [$51.9, $58.0]	$84.4 [$80.7, $88.1]	$16.5 [$15.8, $17.2]
						$128.5 [$120.5, $136.4]	$202.5 [$192.9, $212.1]	$40.6 [$38.9, $42.4]

APPENDIX A

For heart disease, the potential impacts are large, as seen in Table A-8. The life expectancy differential aggregates to $159 billion, disability-free life expectancy to $278.5 billion, and wages aggregate to $30.9 billion. Note that these are driven in part due to higher lifetime risk for non-Hispanic white females. The impacts for the other groups are similar in size to the diabetes scenario. Wage effects are smaller for heart disease than for diabetes due to later onset of heart disease.

Narrowing the gap in hypertension's impact on these populations also shows significant potential for value. In aggregate, the life expectancy gains are valued at $217.4 billion. Disability-free life expectancy gains are valued at $442.1 billion. Wage impacts total $42.2 billion.

Valuing the Potential Gains for the Future Elderly Population

Finally, expanding beyond the narrow birth cohort considered above, we assessed the potential for innovation by looking at the U.S. population of underrepresented individuals over the age of 50 through 2050. The approach is comparable to the cohort results, but now incorporates all individuals 51 and older through 2050 and values the potential for narrowing disparities. These results are presented in Table A-10.

The combination of a large number of aging individuals, high lifetime risk, and large disparities aggregates to sizable potential gains. The estimated potential in diabetes is $2.8 trillion for life expectancy, $4.3 trillion for disability-free life, and $800 billion in years of work. Heart disease aggregates to $3.5 trillion in life expectancy, $5.8 trillion in disability-free life, and $500 billion in years of work. Hypertension is the largest in longevity-related measures, with $4.8 trillion in life expectancy and $9.4 trillion in disability-free life, with $700 billion in years of work.

Discussion

The reduced-form estimates of the differential impact of disease on lesser-represented groups in clinical trials translate into large impacts for individuals who are projected to develop those diseases. Across the diseases, life expectancy impacts range from 0.8 to 1.1 years. Disability-free life expectancy impacts are larger, ranging from 1.0 to 1.7 years. The impact on workforce participation ranges from 0.2 to 0.6 years. When valued in aggregate across all individuals affected in the 51–52-year-old cohort, the potential value is large, ranging from tens to hundreds of billions of dollars. Critically, this is only for one particular cohort of individuals, so the societal value across additional cohorts is even larger.

When aggregated to the over-50 population through 2050, the societal value is sizable.

TABLE A-8 Aggregate Value of Heart Disease Scenario

	N	Lifetime heart disease risk	LE (discounted)	DFLY (discounted)	Work years (discounted)	Aggregate LE	Aggregate DFLY	Aggregate WY
Hispanic females	633,641	70%	0.30 [0.29, 0.32]	0.51 [0.50, 0.52]	0.12 [0.12, 0.13]	$20.3 [$19.4, $21.2]	$34.0 [$33.1, $34.9]	$2.7 [$2.7, $2.8]
Hispanic males	648,817	68%	0.28 [0.26, 0.29]	0.48 [0.46, 0.49]	0.13 [0.13, 0.14]	$18.2 [$17.3, $19.1]	$31.3 [$30.5, $32.2]	$3.0 [$2.9, $3.0]
Non-Hispanic Black females	576,820	57%	0.31 [0.29, 0.32]	0.51 [0.50, 0.52]	0.15 [0.15, 0.16]	$15.1 [$14.4, $15.8]	$25.0 [$24.3, $25.7]	$2.5 [$2.4, $2.5]
Non-Hispanic Black males	509,836	62%	0.35 [0.33, 0.36]	0.57 [0.55, 0.58]	0.21 [0.20, 0.21]	$16.3 [$15.5, $17.0]	$26.8 [$26.1, $27.5]	$3.2 [$3.2, $3.3]
Non-Hispanic white females	2,920,961	61%	0.33 [0.32, 0.35]	0.60 [0.59, 0.62]	0.22 [0.21, 0.23]	$89.2 [$84.9, $93.4]	$161.3 [$156.9, $165.6]	$19.5 [$19.0, $20.1]
						$159.0 [$151.5, $166.6]	$278.5 [$270.9, $286.0]	$30.9 [$30.0, $31.8]

TABLE A-9 Aggregate Value of Hypertension Scenario

	N	Lifetime hypertension risk	LE (discounted)	DFLY (discounted)	Work years (discounted)	Aggregate LE	Aggregate DFLY	Aggregate WY
Hispanic females	633,641	86%	0.28 [0.26, 0.30]	0.66 [0.64, 0.68]	0.15 [0.14, 0.15]	$23.1 [$21.6, $24.5]	$54.3 [$52.5, $56.0]	$4.0 [$3.8, $4.2]
Hispanic males	648,817	88%	0.31 [0.29, 0.33]	0.64 [0.62, 0.66]	0.18 [0.17, 0.19]	$26.6 [$24.9, $28.6]	$54.7 [$53.0, $56.5]	$5.1 [$4.9, $5.4]
Non-Hispanic Black females	576,820	93%	0.36 [0.31, 0.41]	0.73 [0.68, 0.78]	0.21 [0.19, 0.23]	$29.1 [$25.3, $32.9]	$58.7 [$54.9, $62.8]	$5.7 [$5.2, $6.1]
Non-Hispanic Black males	509,836	95%	0.36 [0.33, 0.39]	0.69 [0.67, 0.72]	0.23 [0.22, 0.24]	$26.3 [$24.0, $28.5]	$50.0 [$48.1, $51.9]	$5.5 [$5.2, $5.8]
Non-Hispanic white females	2,920,961	93%	0.28 [0.26, 0.30]	0.55 [0.54, 0.57]	0.16 [0.15, 0.17]	$112.3 [$104.9, $119.8]	$224.3 [$217.1, $231.4]	$21.9 [$20.8, $23.0]
						$217.4 [$200.5, $234.2]	$442.1 [$425.4, $458.7]	$42.2 [$39.9, $44.6]

TABLE A-10 Population Value for Scenarios through 2050

Disease	N	Lifetime risk	LE (discounted)	DFLY (discounted)	Work Years (discounted)	Aggregate LE ($T)	Aggregate DFLY ($T)	Aggregate WY ($T)
Diabetes	161,500,000	57%	0.20 [0.17, 0.23]	0.31 [0.28, 0.35]	0.17 [0.15, 0.19]	$2.8 [2.4, 3.2]	$4.3 [3.8, 4.8]	$.8 [0.7, 0.9]
Heart disease	161,500,000	64%	0.23 [0.20, 0.25]	0.37 [0.35, 0.40]	0.09 [0.09, 0.10]	$3.5 [3.2, 3.9]	$5.8 [5.4, 6.2]	$.5 [0.5, 0.5]
Hypertension	161,500,000	91%	0.22 [0.19, 0.26]	0.43 [0.39, 0.46]	0.10 [0.09, 0.11]	$4.8 [4.1, 5.6]	$9.4 [8.6, 10.1]	$.7 [0.6, 0.8]
						$11.2 [9.6, 12.7]	$19.5 [17.9, 21.2]	$2.0 [1.8, 2.2]

Limitations

This type of analysis is subject to many limitations. A key assumption is that the transition models estimated using the HRS data will hold into the future. A reduced-form approach to modeling likely leaves out important factors, loading the estimated effect onto a particular variable.

Transition Model Estimates

Diabetes includes the transition model estimates for 2-year mortality, disability, and working for pay, as well as the marginal effects for diabetes (see Table A-11). The key parameter of interest, "underrepresented and has diabetes," has a 0.6 percentage point increase in 2-year mortality, a 2.8 percentage point increase in reporting disability, and a 3.3 percentage point reduction in working for pay.

Similarly, Table A-12 shows the transition models for key outcomes in the heart disease analysis. Here, the key parameter of interest, "underrepresented and has heart disease," is associated with a 0.9 percentage point increase in 2-year mortality, a 5.6 percentage point increase in reporting disability, and a 3.8 percentage point reduction in working for pay.

Finally, hypertension shows comparable estimates for the hypertension analysis (see Table A-13). In this specification, "underrepresented and has hypertension" is associated with a 0.6 percentage point increase in 2-year mortality, a 3.5 percentage point increase in reporting disability, and a 1.4 percentage point decrease in working for pay.

TABLE A-11 Diabetes

	Mortality	Margins	Disability	Margins	Work	Margins
	b	b	b	b	b	b
Main						
2-year lag of diabetes ever	0.288***	0.034***	0.323***	0.093***	-0.224***	-0.062***
Underrepresented and has diabetes	0.058*	0.006*	0.100***	0.028***	-0.118***	-0.033***
White males	0	0	0	0	0	0
Black males	0.412	0.012***	-0.327	0.057***	-2.045***	-0.077***
Hispanic males	0.136	-0.009**	-0.459	0.059***	-1.962***	-0.042***
White females	0.227	-0.020***	-0.21	0.013***	-1.926***	-0.096***
Black females	0.81	-0.014***	-0.299	0.117***	-3.134***	-0.118***
Hispanic females	1.012	-0.032***	-1.434***	0.111***	-3.511***	-0.161***
Age spline under 65	0.037***	0.003***	-0.002	0.001***	-0.089***	-0.019***

continued

TABLE A-11 Continued

	Mortality	Margins	Disability	Margins	Work	Margins
	b	b	b	b	b	b
Main						
Age spline 65–74	0.036***	0.004***	0.026***	0.005***	-0.077***	-0.027***
Age spline 75–84	0.052***	0.006***	0.056***	0.016***	-0.062***	-0.020***
Age spline over 85	0.083***	0.008***	0.067***	0.019***	-0.086***	-0.021***
Black males # age spline under 65	-0.004		0.010*		0.029***	
Hispanic males # age spline under 65	-0.005		0.012*		0.033***	
White females # age spline under 65	-0.007		0.004		0.028***	
Black females # age spline under 65	-0.013		0.012***		0.046***	
Hispanic females # age spline under 65	-0.022*		0.032***		0.052***	
Black males # age spline 65–74	-0.002		-0.019***		0.015**	
Hispanic males # age spline 65–74	0.012		-0.006		-0.043***	
White females # age spline 65–74	0.002		-0.003		-0.024***	
Black females # age spline 65–74	-0.012		-0.015***		-0.019***	
Hispanic females # age spline 65–74	0.002		-0.022***		-0.065***	
Black males # age spline 75–84	-0.002		0.01		-0.027*	
Hispanic males # age spline 75–84	0.008		0.004		-0.048**	
White females # age spline 75–84	0		0.006		-0.004	
Black females # age spline 75–84	0.003		0.016**		-0.012	
Hispanic females # age spline 75–84	0.01		0.002		-0.004	
Black males # age spline over 85	-0.016		-0.041**		0.115***	
Hispanic males # age spline over 85	-0.024		-0.036*		-0.055	

TABLE A-11 Continued

	Mortality	Margins	Disability	Margins	Work	Margins
	b	b	b	b	b	b
Main						
White females # age spline over 85	-0.002		0.022***		0.003	
Black females # age spline over 85	-0.012		-0.016		0.004	
Hispanic females # age spline over 85	-0.013		0.008		0.06	
Constant	-4.232***		-1.151***		5.678***	
r2_p	0.16		0.089		0.23	
N	191,036	191,036	178,803	178,803	166,827	166,827

NOTE: Asterisks represent statistical significance. ***p<0.001, ** p<0.01, * p<0.05.

TABLE A-12 Heart Disease

	Mortality	Margins	Disability	Margins	Work	Margins
	b	b	b	b	b	b
Main						
Lag of heart disease ever	0.355***	0.041***	0.277***	0.079***	-0.261***	-0.073***
Underrepresented and has heart disease	0.087***	0.009***	0.197***	0.056***	-0.135***	-0.038***
White males	0	0	0	0	0	0
Black males	0.26	0.019***	-0.466	0.066***	-1.940***	-0.089***
Hispanic males	0.227	0	-0.381	0.075***	-2.036***	-0.057***
White females	0.111	-0.019***	-0.242	0.008**	-1.881***	-0.098***
Black females	0.65	-0.007**	-0.429*	0.128***	-3.021***	-0.130***
Hispanic females	0.981	-0.022***	-1.536***	0.139***	-3.407***	-0.178***
Age spline under 65	0.034***	0.003***	-0.003	0.001***	-0.088***	-0.019***
Age spline 65–74	0.032***	0.003***	0.024***	0.005***	-0.075***	-0.026***
Age spline 75–84	0.047***	0.005***	0.052***	0.015***	-0.058***	-0.019***
Age spline over 85	0.081***	0.008***	0.066***	0.018***	-0.085***	-0.020***
Black males # age spline under 65	-0.001		0.013**		0.026***	
Hispanic males # age spline under 65	-0.006		0.011*		0.033***	

continued

TABLE A-12 Continued

	Mortality	Margins	Disability	Margins	Work	Margins
	b	b	b	b	b	b
Main						
White females # age spline under 65	-0.005		0.004		0.027***	
Black females # age spline under 65	-0.01		0.015***		0.044***	
Hispanic females # age spline under 65	-0.021*		0.035***		0.049***	
Black males # age spline 65–74	-0.004		-0.022***		0.016**	
Hispanic males # age spline 65–74	0.014		-0.004		-0.043***	
White females # age spline 65–74	0.001		-0.004		-0.024***	
Black females # age spline 65–74	-0.01		-0.013**		-0.020***	
Hispanic females # age spline 65–74	0.008		-0.018***		-0.066***	
Black males # age spline 75–84	0.001		0.011		-0.029**	
Hispanic males # age spline 75–84	0.01		0.005		-0.052**	
White females # age spline 75–84	0		0.006		-0.005	
Black females # age spline 75–84	0.002		0.013*		-0.011	
Hispanic females # age spline 75–84	0.009		0		-0.006	
Black males # age spline over 85	-0.02		-0.048***		0.122***	
Hispanic males # age spline over 85	-0.024		-0.037*		-0.046	
White females # age spline over 85	-0.004		0.020***		0.004	
Black females # age spline over 85	-0.016*		-0.020*		0.007	
Hispanic females # age spline over 85	-0.016		0.004		0.067	
Constant	-4.113***		-1.097***		5.620***	

TABLE A-12 Continued

	Mortality	Margins	Disability	Margins	Work	Margins
	b	b	b	b	b	b
Main						
r2_p	0.168		0.091		0.231	
N	191055	191055	178824	178824	166848	166848

NOTE: Asterisks represent statistical significance. ***$p<0.001$, ** $p<0.01$, * $p<0.05$.

TABLE A-13 Hypertension

	Mortality	Margins	Disability	Margins	Work	Margins
	b	b	b	b	b	b
Main						
Lag of hypertension ever	0.183***	0.019***	0.172***	0.046***	-0.219***	-0.063***
Underrepresented and has hypertension	0.057**	0.006**	0.129***	0.035***	-0.050**	-0.014**
White males	0	0	0	0	0	0
Black males	0.336	0.009**	-0.379	0.041***	-1.976***	-0.071***
Hispanic males	0.234	-0.008*	-0.363	0.057***	-2.053***	-0.048***
White females	0.336	-0.025***	-0.094	-0.005	-2.010***	-0.093***
Black females	0.744	-0.017***	-0.327	0.097***	-3.121***	-0.108***
Hispanic females	1.064*	-0.033***	-1.391***	0.110***	-3.544***	-0.166***
Age spline under 65	0.037***	0.003***	-0.002	0.001***	-0.088***	-0.018***
Age spline 65–74	0.036***	0.004***	0.026***	0.005***	-0.077***	-0.026***
Age spline 75–84	0.052***	0.006***	0.055***	0.016***	-0.062***	-0.019***
Age spline over 85	0.082***	0.008***	0.067***	0.018***	-0.086***	-0.021***
Black males # age spline under 65	-0.003		0.010*		0.028***	
Hispanic males # age spline under 65	-0.006		0.010*		0.034***	
White females # age spline under 65	-0.009		0.001		0.029***	
Black females # age spline under 65	-0.013		0.012**		0.047***	
Hispanic females # age spline under 65	-0.023*		0.031***		0.052***	
Black males # age spline 65–74	-0.003		-0.021***		0.016**	

TABLE A-13 Continued

	Mortality	Margins	Disability	Margins	Work	Margins
	b	b	b	b	b	b
Main						
Hispanic males # age spline 65–74	0.013		-0.007		-0.041***	
White females # age spline 65–74	0.001		-0.006		-0.022***	
Black females # age spline 65–74	-0.011		-0.013**		-0.020***	
Hispanic females # age spline 65–74	0.002		-0.023***		-0.063***	
Black males # age spline 75–84	-0.001		0.01		-0.026*	
Hispanic males # age spline 75–84	0.005		0.003		-0.049**	
White females # age spline 75–84	-0.001		0.004		-0.002	
Black females # age spline 75–84	0		0.012*		-0.008	
Hispanic females # age spline 75–84	0.007		-0.001		-0.003	
Black males # age spline over 85	-0.017		-0.044**		0.112***	
Hispanic males # age spline over 85	-0.025		-0.037*		-0.046	
White females # age spline over 85	-0.004		0.021***		0.003	
Black females # age spline over 85	-0.015*		-0.019*		0.007	
Hispanic females # age spline over 85	-0.014		0.006		0.062	
Constant	-4.289***		-1.188***		5.683***	
r2_p	0.156		0.085		0.231	
N	191014	191014	178786	178786	166815	166815

NOTE: Asterisks represent statistical significance. ***p<0.001, ** p<0.01, * p<0.05.

REFERENCES

Goldman, D., P.-C. Michaud, D. Lakdawalla, Y. Zheng, A. Gailey, and I. Vaynman. (2010). The fiscal consequences of trends in population health. *National Tax Journal* 63(2):307.

Goldman, D. P., B. Shang, J. Bhattacharya, A. M. Garber, M. Hurd, G. F. Joyce, D. Lakdawalla, C. Panis, and P. G. Shekelle. 2005. Consequences Of Health Trends And Medical Innovation For The Future Elderly: When demographic trends temper the optimism of biomedical advances, how will tomorrow's elderly fare? *Health Affairs* 24(Suppl. 2):W5-R5–W5-R17.

Goldman, D. P., Y. Zheng, F. Girosi, P.-C. Michaud, S. J. Olshansky, D. Cutler, and J. W. Rowe. 2009. The benefits of risk factor prevention in Americans aged 51 years and older. *American Journal of Public Health* 99(11):2096–2101.

Lakdawalla, D. N., D. P. Goldman, and B. Shang. 2005. The Health and Cost Consequences Of Obesity Among The Future Elderly. *Health Affairs* 24(Suppl. 2), W5-R30-W35-R41.

Leaf, D. E., B. Tysinger, D. P. Goldman, and D. N. Lakdawalla. 2020. Predicting quantity and quality of life with the Future Elderly Model. *Health Economics*.

NASEM (National Academies of Sciences, Engineering, and Medicine). 2015. *The Growing Gap in Life Expectancy by Income: Implications for Federal Programs and Policy Responses*. Washington, DC: The National Academies Press.

RAND HRS (Health and Retirement Study). 2021a. HRS public use dataset. Produced and distributed by the University of Michigan.

RAND HRS. 2021b. HRS public use dataset. Produced by the RAND Center for the Study of Aging.

Van Nuys, K. E., Z. Xie, B. Tysinger, M. A. Hlatky, and D. P. Goldman. 2018. Innovation in Heart Failure Treatment: Life Expectancy, Disability, and Health Disparities. *JACC. Heart failure*. doi:10.1016/j.jchf.2017.12.006

Zissimopoulos, J. M., B. C. Tysinger, P. A. St. Clair, and E. M. Crimmins. 2018. The impact of changes in population health and mortality on future prevalence of Alzheimer's disease and other dementias in the United States. *The Journals of Gerontology: Series B*, 73(Suppl. 1):S38–S47.

Appendix B

Key Trends in Demographic Diversity in Clinical Trials

Jakub P. Hlávka[1]

BACKGROUND

Insufficient demographic diversity in clinical trials has long been recognized as an issue that may hinder innovation and access to therapies. In the past three decades, however, diversity in clinical trials became a policy priority, advanced by federal agencies such as the Food and Drug Administration (FDA) Office of Women's Health and the Society for Women's Health Research (SWHR), and later by the FDA Office of Minority Health (OMH), established in 2010 (FDA, 2011). In 1992, the General Accounting Office (GAO)—the historical predecessor to the Government Accountability Office—released the report *Women's Health: FDA Needs to Ensure More Study of Gender Differences in Prescription Drug Testing*, ushering a new era of focus on the issue of diversity in clinical trials.

Soon after, the National Institutes of Health Revitalization Act of 1993 followed, which directed the National Institutes of Health (NIH) to establish guidelines for inclusion of women and minorities in clinical research (P.L. 103-43). Since then, multiple guidelines and regulatory documents have addressed the issue, including a 1993 guidance on the study and evaluation of gender difference in clinical trial evaluation of drugs, which lifted restriction on participation by most women with childbearing potential (FDA, 1993, 2020b). In 2008, the ClinicalTrials.gov results database was launched to implement Section 801 of the FDA Amendments Act of 2007, or FDAAA 801 (P.L. 110–85), which requires the submission of "basic results" for applicable clinical trials (ACT) no later than 12

---

[1] Available at jakub.hlavka@usc.edu.

months after their primary completion date.[2,3] The submission of adverse event information has been required since September 2009. Basic results are defined as (a) participant flow, (b) baseline characteristics, (c) outcome measures and statistical analyses, and (d) adverse events.[4]

More recently, the FDA Reauthorization Act of 2017 (FDARA) made a reference encouraging the "enrollment of more diverse patient populations," and the Final Rule for Clinical Trials Registration and Results Information Submission (42 CFR Part 11) went into effect in January 2017.[5] The rule requires the submission of "baseline or demographic characteristic measured in the clinical trial, including age, sex/gender, race, ethnicity (if collected under the protocol)" for trials that are required to be registered under Section 11.22 (Phase 1 trials are excluded),[6] for reporting purposes via ClinicalTrials.gov.[7] A 2020 FDA *Guidance for Industry* (FDA, 2020) made the following (nonbinding) calls to action focusing on[8]

- broadening eligibility criteria and avoiding unnecessary exclusions for clinical trials;
- developing eligibility criteria and improving trial recruitment so that the participants enrolled in trials will better reflect the population most likely to use the drug, if the drug is approved, while maintaining safety and effectiveness standards; and
- applying the recommendations for broadening eligibility criteria to clinical trials of drugs intended to treat rare diseases or conditions.

In November 2020, the FDA issued its guidance on "Enhancing the Diversity of Clinical Trial Populations," which considers both demographic characteristics of study populations (e.g., sex, race, ethnicity, age, location of residency) and non-demographic characteristics of populations (e.g., patients with organ dysfunction, comorbid conditions, disabilities, those at the extremes of the weight range, and populations with diseases or conditions with low prevalence) (FDA,

---

[2] See https://clinicaltrials.gov/ct2/about-site/results.

[3] Applicable clinical trials "generally include interventional studies (with one or more arms) of FDA-regulated drug, biological, or device products that meet one of the following conditions: a) The trial has one or more sites in the United States; b) The trial is conducted under an FDA investigational new drug application or investigational device exemption; c) The trial involves a drug, biological, or device product that is manufactured in the United States or its territories and is exported for research."

[4] See https://clinicaltrials.gov/ct2/about-site/results.

[5] See https://clinicaltrials.gov/ct2/manage-recs/fdaaa.

[6] See https://ecfr.federalregister.gov/current/title-42/chapter-I/subchapter-A/part-11/subpart-B/section-11.22.

[7] See https://ecfr.federalregister.gov/current/title-42/chapter-I/subchapter-A/part-11.

[8] It also referenced other policy documents, including the 1993 International Conference on Harmonization (ICH) guideline, which advised against arbitrary upper-age cutoffs in clinical trials that may result in underrepresentation of older adults. In that document, ICH representatives recognized important pharmacokinetic differences between younger and older patients related to renal and hepatic function, as well as drug-drug interactions.

2020b). This guidance also describes "enrichment strategies"—targeted inclusion of certain populations with the goal to more readily demonstrate a drug effect—and recommends that even with enrichment, trials should keep their enrollment criteria as broad and representative as possible.

The guidance also listed approaches that can improve diversity of enrolled participants, including

- making trial participation less burdensome for participants;
- adopting enrollment and retention practices that enhance inclusiveness, such as public outreach, education, community engagement, include varied geographic locations, offer multilingual resources, use real-world data, leverage social media, etc.; and
- offering expanded access—to diagnose, monitor, or treat a patient's condition as part of a clinical trial (this may help identify patients for subsequent studies).

This Appendix offers a brief overview of trends in diverse enrollment and data reporting across clinical trials, and references lessons-learned and insights from other research related to demographic diversity in trial enrollment.

PAST EVIDENCE

Historically, data on population demographics across clinical trials have not been consistently reported, particularly prior to the 2017 guidance, which requires reporting by sex/gender, race, and/or ethnicity in applicable clinical trials. Evidence so far has emerged in multiple individual research reports on different aspects of diversity in clinical trials.

Reporting of Demographic Data

Demographic data of enrolled participants in clinical trials in the United States is most transparently collected by ClinicalTrials.gov, a registry maintained by the National Library of Medicine at the NIH. Additional reporting on NIH-funded research (intra- and extramural) has recently also been reported by NIH institutes and centers. In both cases, limitations to how often and comprehensively data are made available have made longitudinal institute-level data difficult to examine. As we show below, not all trials report their demographic characteristics—further work to improve the level and quality of reporting is still needed.

Past Research on Gender Diversity

Despite the regulatory efforts to increase gender diversity in trial enrollment, some have suggested limited progress has been made (Clark et al., 2019). This is particularly significant given that sex differences are observed in response to some drugs, including the prevalence of adverse events (Anderson, 2005; FDA, 2011). Evidence from the 1990s and early-2000s suggested relative underrepresentation of women and ethnic minorities in clinical trials (Mak et al., 2007).

A 2012 study has documented an increase in the reporting of trial characteristics in ClinicalTrials.gov in interventional trials from 2007 to 2010 (Califf et al., 2012). Geographic differences and therapeutic areas were linked to diverse trends in gender- and age-specific enrollment.

More recent work has confirmed the challenge of enrolling women in some therapeutic areas: in stroke clinical trials, for instance, women have been underrepresented even after incidence and prevalence of the disease is taken into account (Carcel, 2021), with highest underrepresentation reported in secondary prevention trials (10 percent in one study) (Strong et al., 2020).

However, it appears that the trend has been improving in some areas. For example, somewhat optimistic results were described in a subset of pivotal trials (Phase 2 and 3 trials in support of drug/biologic approval) studied by Eshera et al. (2015): in studies of drugs approved between 2010 and 2012, just 45 percent of trial participants in small molecule trials were women, but they represented 65 percent of participants in biologic trials (based on Drugs@FDA data). The authors concluded that 82 percent of trials had a study population representative of the sex distribution in the intended patient population, but that minority groups still had lower participation rates than would be representative (with 77 percent of participants white, population average 72 percent) (Eshera et al., 2015).

Non-Gender Diversity Measures

Numerous studies have focused on the reporting of diversity in non-gender domains, such as the age of participants (relative to prevalent disease populations) and the reporting by ethnic or racial groups. Highlighted here are several specific studies that provide illustrative evidence of underrepresentation of specific groups in clinical trials in the past two decades.

In 2003, a study of 495 cancer trials between 1997 and 2000 indicated that elderly participants comprised 32 percent of participants in Phase 2 or 3 clinical trials, compared with 61 percent of patients with incident cancers in the United States who are elderly (Lewis et al., 2003). This was more pronounced in trials for early-stage cancers, with protocol exclusion criteria on the basis of organ-system abnormalities and functional status limitations being associated with lower elderly participation (the authors estimate that relaxing those eligibility restrictions would increase their share to 60 percent) (Lewis et al., 2003).

A 2004 analysis of cancer clinical trials made similar conclusions: it found a higher enrollment fraction (relative to incident cases) in younger cohorts—3.0 percent for patients 30–64, 1.3 percent for 65–74, and 0.5 percent for patients 75 and older (Murthy et al., 2004). Hispanic and Black patients were 28 percent and

29 percent less likely to enroll than white patients after adjustment for incidence, age, and other factors. The difference was largest in lung cancer, where Black patients were 39 percent and Hispanic patients 53 percent less likely to enroll than white patients.

Another study of oncology trials conducted between 1994 and 2015 calculated the difference in the age of trial participants and the disease population they studied (Ludmir et al., 2019). For most diseases, trial populations were younger than the population median age, with an average difference of 6.49 years (highest in lung cancer—8.98 years), and higher among industry-funded trials.

A recent study of 230 vaccine trials from 2011 to 2020 indicated that white participants tend to be overrepresented, while Black and other minorities tend to be underrepresented. The enrollment of Asian individuals was similar to U.S. Census estimates (Flores et al., 2021). Only about 12.1 percent of participants in vaccine trials were over 65 years of age. A report on the diversity of mRNA vaccine trials for COVID-19 by the Kaiser Family Foundation has found a relatively higher share of white participants in both trials compared with the U.S. population, resulting in relative underrepresentation of Black and Asian participants. However, the participation of Hispanics exceeded the share of Hispanics in the U.S. population (Artiga et al., 2021). These results, however, originate from trial sites within and outside of the United States (notably Europe and Latin America), which may explain some of the relative overrepresentation of white participants.

Even more recently completed trials have struggled with diverse enrollment—a Phase 2 trial of crenezumab in Alzheimer's disease with 360 participants across 83 sites in 6 countries reported 97.5 percent of participants being white, and only 2.8 percent of all participants being Hispanic, for example (however, women consisted of 55.3 percent of all participants).

Barriers to Diversity

Participation of older adults in cancer clinical trials has not changed over time (Sedrak et al., 2021). Numerous studies have been published on barriers to clinical trial participation by underrepresented groups. Participation of older adults in cancer trials has often been identified as hindered by eligibility criteria, concern for toxicity, concern for patient age, patient knowledge and transportation limitations, and others such as time/burden issues and willingness to participate (Sedrak et al., 2021). Specific barriers that participants face may include (1) mistrust, (2) lack of comfort with the clinical trial process, (3) lack of information about clinical trials, (4) time and resource constraints associated with participation, and (5) lack of awareness about the existence and importance of clinical trials (Clark et al., 2019).

In the mental health area, a review has identified additional challenges, including transportation difficulties, distrust and suspicion of researchers, and stigma attached to mental illness (Woodall et al., 2010). The review included an Alzheimer's disease case study that showed increased participation by Black patients by more than 100 percent when educational strategies and compensation for travel costs were introduced in a trial. Another study of key barriers in Alzheimer's disease research has indicated that enrollment challenges result in slower and more expensive trials, suggesting the need for nationally coordinated efforts to increase diverse participation in clinical trials in this and other disease areas (Malzbender et al., 2020).

RECENT EVIDENCE

Demographics of Trials Resulting in FDA Approval

Participation of Females

We find that the positive trend of increasing representation of females in trials that have resulted in FDA approval has been relatively consistent. Among drugs that have been approved in recent years, the average share of females in the trial population has been consistently reported since 2014 by the FDA, with the average share of 51 percent between 2014 and 2021, ranging from 37 percent in 2014 (six trials) to 57.1 percent in 2019. We analyze 290 approvals for which demographic data are reported in the FDA Snapshots between 2014 and 2021 (as of May 1, 2021). While the mean (unweighted) representation of females achieving an average of 51 percent between 2014 and 2021, ranging from 37 percent in 2014 (six trials) to 54.8 percent in 2020 (data for 2021 are partial only). Prior to 2021, females represented greater than 50 percent of trial participants over at least 5 years in the areas of ophthalmology, gastroenterology, and endocrinology/metabolism/bone. In turn, men have represented greater than 50 percent of trial participants over at least 5 years in the areas of cardiovascular disease and infectious disease (viral). Representation of females across all trials and by therapeutic area is indicated in Figure B-1.

We also find women to be relatively more represented in non-oncology trials between 2014 and 2021 (non-gender-specific trials only), as shown in Figure B-2.

Participation of White Patients

Among approved drugs, participation of white patients has ranged from 84 percent in 2014 to 73.7 percent in 2020, indicating a relatively consistent decrease in the share of white participants in trials resulting in FDA approval during this period (2021 data are yet incomplete). Figure B-3 shows trends by approvals and therapeutic areas in. Prior to 2021, white patients represented greater than 70 percent of trial participants over at least 5 years in the areas of analgesia/anesthesiology/anti-inflammatory, cardiovascular disease, endocrinology/metabolism/bone, infectious disease (nonviral), oncology, ophthalmology, and pulmonary.

APPENDIX B

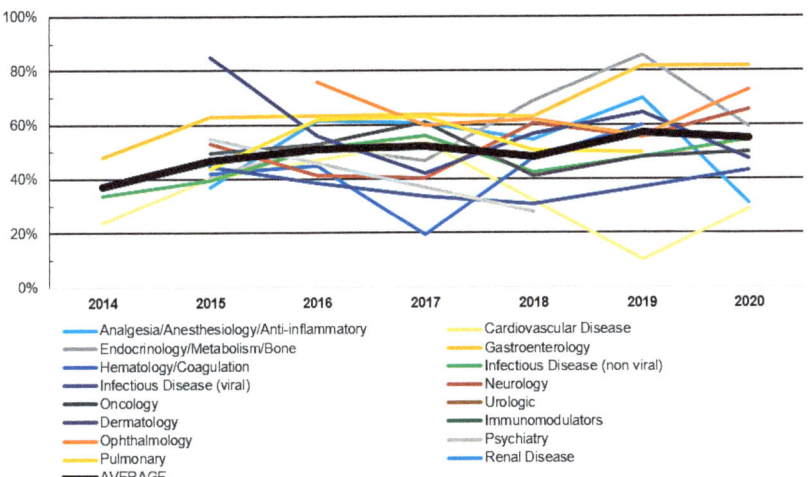

FIGURE B-1 Average % of females in trials by year of FDA approval and therapeutic area (n = 287).
SOURCE: Analysis of FDA Drug Trials Snapshots as of May 2021.

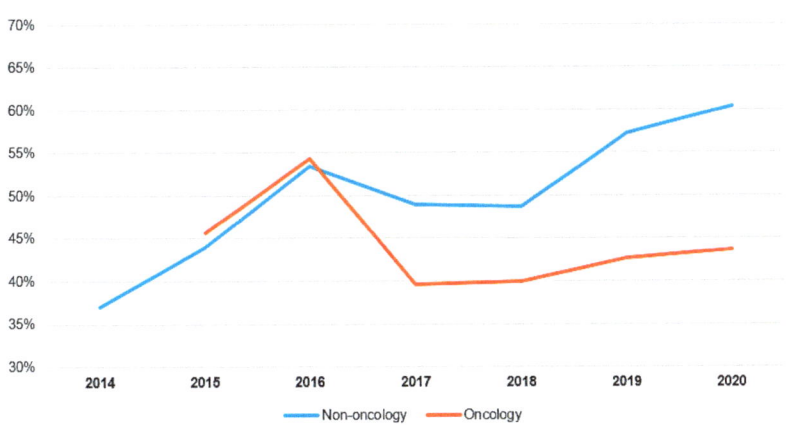

FIGURE B-2 Mean % of females by year of FDA approval (non-gender-specific trials only, n = 255).

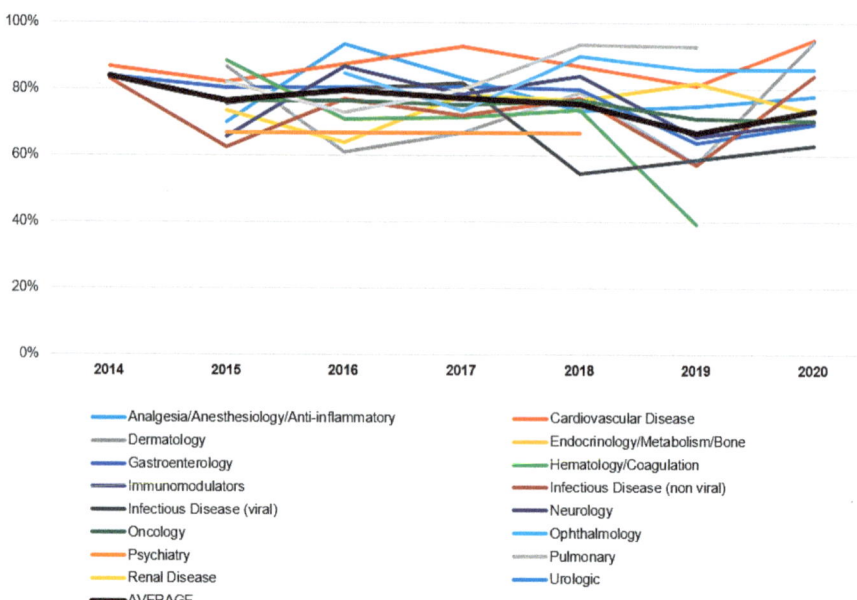

FIGURE B-3 Average % of white patients in trials by year of FDA approval and therapeutic area (n = 287).

Participation of the Elderly

Among approved drugs, participation of patients over 65 has ranged from 10 percent in 2014 to 39.9 percent in 2020, indicating a consistent increase in the share of elderly participants in trials resulting in FDA approval during this period (2021 data are yet incomplete). Figure B-4 shows trends by approvals and therapeutic areas. Prior to 2021, elderly patients represented at least 25 percent of trial participants over at least 5 years in the areas of cardiovascular disease, neurology, and oncology. In none of the years for which reporting is available, elderly patients represented greater than 25 percent of trial participants in the areas of gastroenterology, renal disease, psychiatry, and dermatology.

Demographics of Trials Funded by NIH Institutes/Centers

A review of participation in clinical research funded by different NIH institutes and centers revealed that participation by females has been steadily increasing over 2013–2018 for which data are available (no data were reported in 2015, but reporting requirements changed in FY 2016, resulting in an increase in

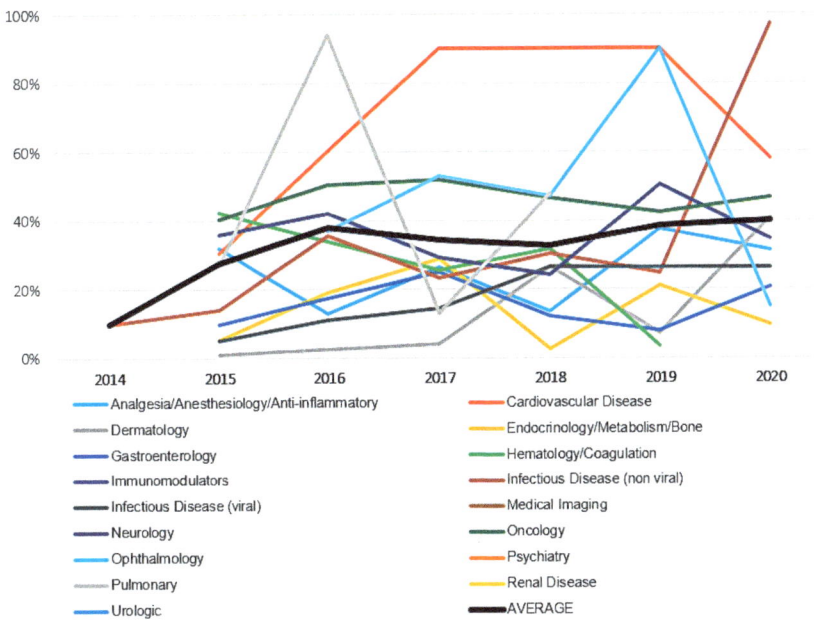

FIGURE B-4 Average % of patients over 65 in trials by year of FDA approval and therapeutic area (n = 287).

participants reported across NIH institutes and centers).[9] Across all NIH institutes and centers, mean representation of females in trials was 44.3 percent in 2013, 47.2 percent in 2014, 54.1 percent in 2016, 47.9 percent in 2017, and 52.4 percent in 2018 (on average 22.1 million participants were included in NIH-funded trials during each of these annual reporting periods).

As shown in Figure B-5, among the top 10 largest institutes/centers by trial enrollment (which represent 89.7 percent of enrollment across all institutes/centers), females represented at least 50 percent of participants in trials supported by the National Institute of Environmental Health Sciences, the National Institute of Diabetes and Digestive and Kidney Diseases, and the National Institute of Child Health and Human Development across all years of reporting, and at least 50 percent of participants in at least 3 years of reporting in trials supported by the National Institute on Aging, the Clinical Center, the National Institute of Allergy and Infectious Diseases, and the National Heart, Lung, and Blood

---

[9] Changes included the requirement for career development awards (Ks) and fellowships (Fs) to report inclusion data, and the NIH stopped granting exceptions for use of existing datasets or resources, early-phase feasibility studies, and others.

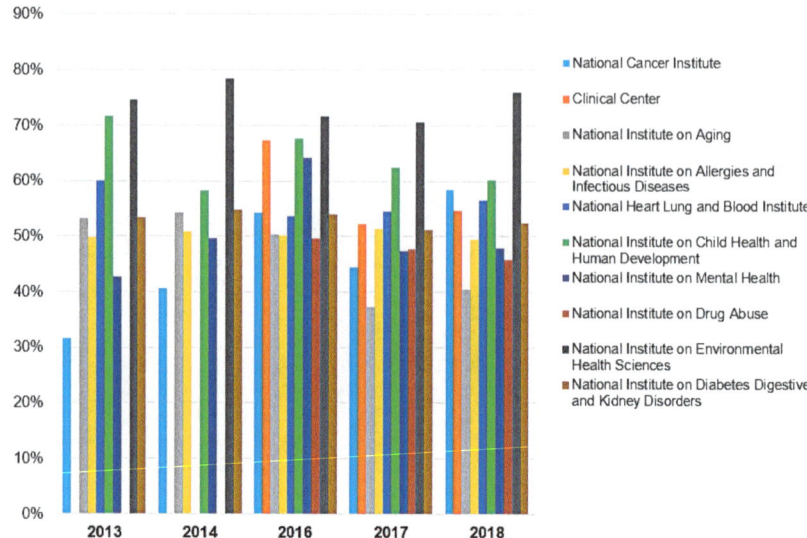

FIGURE B-5 Participation of females in clinical trials supported by NIH institutes (top 10 institutes/centers by 2018 enrollment).

Institute. Across all 5 years of reporting, females never exceeded 50 percent of participants in trials supported by the National Institute on Minority Health and Health Disparities and the National Institute on Alcohol Abuse and Alcoholism.

The reporting on participation by ethnic minorities in NIH reports is less consistent. However, ethnicity data are more available than race demographics data, as shown in Table B-1.

In Figure B-6, the share of white participants in clinical trials sponsored by top NIH institutes is reported. These results show a relatively stable trend, with the weighted average of white participants among the top 10 institutes ranging from 51.8 percent in 2013 to 60.6 percent in 2018 (this trend mirrors that of all NIH-sponsored trials, as shown in Figure B-1). The lowest representation of white participants was in trials sponsored by the National Institute of Allergy and Infectious Diseases and the National Institute of Child Health and Human Development (both consistently less than 50 percent). The U.S. Census estimate of white Americans (alone, not Hispanic or Latino) as a share of the U.S. population was 60.1 percent in 2019 (white alone without ethnicity specification was estimated at 76.3 percent of the U.S. population).[10]

In Figure B-7, the share of African American/Black participants in clinical trials sponsored by top NIH institutes is reported. These results show a relatively stable trend, with the weighted average of African American/Black participants

---

[10] See https://www.census.gov/quickfacts/fact/table/US/PST045219.

APPENDIX B

TABLE B-1 Demographics of Participants in Trials Supported by NIH Centers and Institutes

	2013 (%)	2014 (%)	2016 (%)	2017 (%)	2018 (%)
Female	44.3	47.2	54.1	47.9	52.4
American Indian	2.1	1.3	0.8	0.7	1.0
Asian	15.1	17.2	8.4	26.4	7.8
Black/African American	12.2	14.3	10.0	10.8	13.5
Native Hawaiian/Pacific Islander	0.3	0.3	0.6	0.1	0.2
White	52.9	49.5	49.6	49.9	60.0
More than 1 race	1.1	1.1	2.0	1.9	2.3
Unknown race	1.1	1.1	2.0	1.9	2.3
Hispanic	9.8	8.1	10.8	6.7	8.5
Non-Hispanic	86.1	89.6	62.6	81.8	76.2
Unknown ethnicity	4.1	2.3	22.4	9.8	12.0
Sum of all races	84.7	84.8	73.5	91.8	87.2
Sum of all ethnicities	100.0	100.0	95.8	98.3	96.7

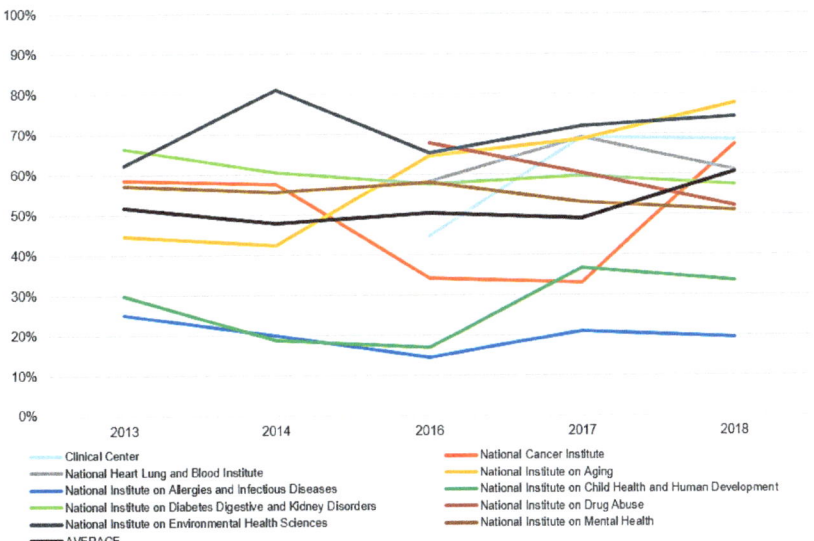

FIGURE B-6 Share of white participants in clinical trials by NIH institutes (top 10 institutes/centers by 2018 enrollment).

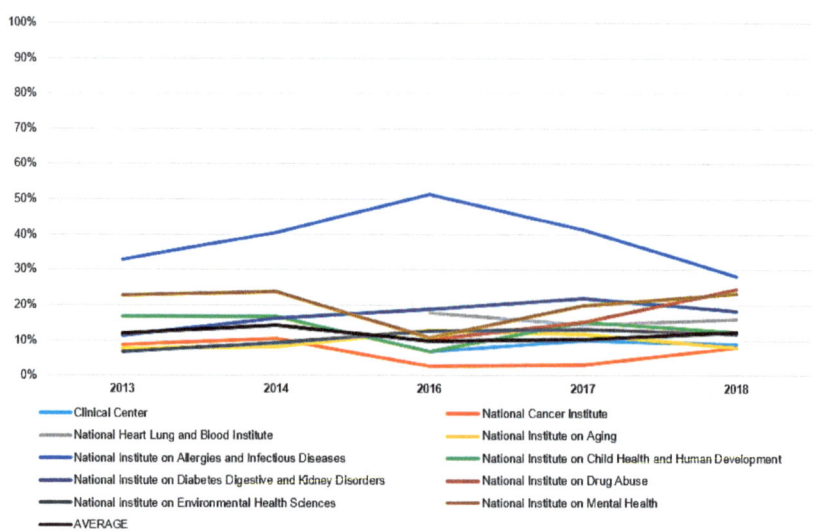

FIGURE B-7 Share of African American/Black participants in clinical trials by NIH institutes (top 10 institutes/centers by 2018 enrollment).

among the top 10 institutes ranging from 11.9 percent in 2013 to 12.3 percent in 2018. The highest representation of African American/Black participants was in trials supported by the National Institute of Allergy and Infectious Diseases (greater than 25 percent in all years). The lowest representation of African American/Black participants was in trials sponsored by the National Cancer Institute (reaching 10.5 percent at most). The U.S. Census estimate of Black or African Americans (alone) as a share of the U.S. population was 13.4 percent in 2019.[11]

In Figure B-8, the share of Asian participants in clinical trials sponsored by top NIH institutes is reported. These results show a less stable trend, with the weighted average of Asian participants among the top 10 institutes ranging from 8.2 percent in 2016 to 27.7 percent in 2018 (driven by large Asian enrollment in National Cancer Institute trials during that year). The highest representation of Asian participants was in trials supported by the National Institute of Allergy and Infectious Diseases (greater than 20 percent on average). The lowest representation of Asian participants was in trials sponsored by the National Institute of Diabetes and Digestive and Kidney Diseases (averaging under 5 percent). The U.S. Census estimate of Asian Americans (alone) as a share of the U.S. population was 5.9 percent in 2019.[12]

In Figure B-9, the share of Hispanic participants in clinical trials sponsored by top NIH institutes is reported. These results show a relatively stable trend,

---

[11] See https://www.census.gov/quickfacts/fact/table/US/PST045219.
[12] See https://www.census.gov/quickfacts/fact/table/US/PST045219.

APPENDIX B

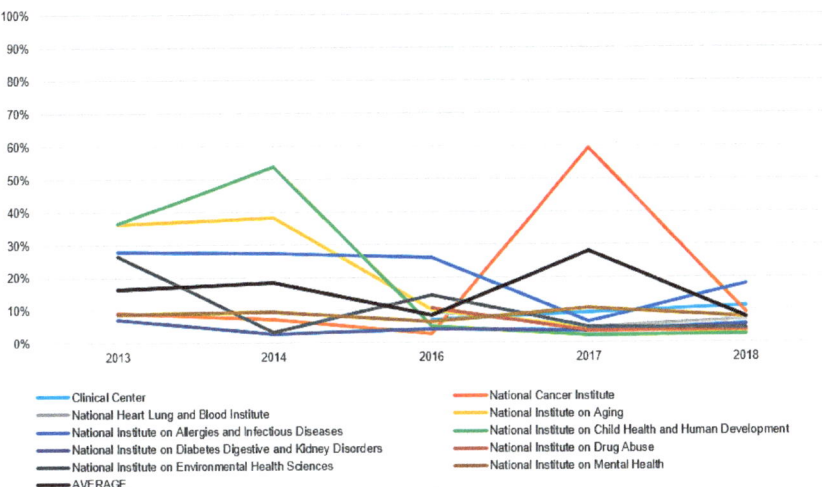

FIGURE B-8 Share of Asian participants in clinical trials by NIH institutes (top 10 institutes/centers by 2018 enrollment).

with the weighted average of Hispanic participants among the top 10 institutes ranging from 6.7 percent in 2017 to 11.3 percent in 2016. The highest representation of Hispanic participants was in trials supported by the National Institute on Drug Abuse (greater than 16 percent on average). The lowest representation of Hispanic participants was in trials sponsored by the National Cancer Institute and

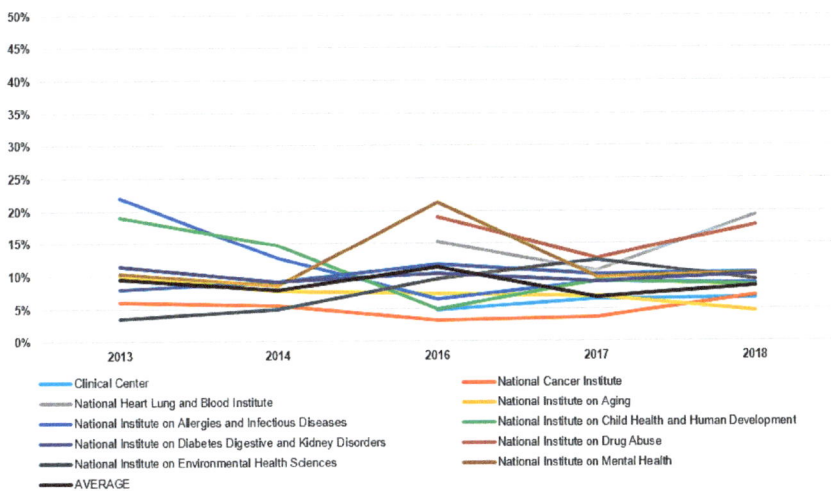

FIGURE B-9 Share of Hispanic participants in clinical trials by NIH institutes (top 10 institutes/centers by 2018 enrollment).

the Clinical Center (both under 6 percent). The U.S. Census estimate of Hispanic Americans as a share of the U.S. population was 18.5 percent in 2019.[13]

In the Supplementary Material section, we report key characteristics of Phase 3 trials only. Of note, female participation in Phase 3 trials has been higher relative to all trials in all reporting periods except for 2016 (the highest difference in relative representation of females was observed in 2013, where Phase 3 trials reported 64.2 percent females, while all trials reported just 44.3 percent females).

Reporting Gaps

As indicated above, data for racial and ethnic and subgroups in clinical trials have not been consistently reported in NIH-supported trials, with just 73.5–91.8 percent of participants reported with their race over the period for which reporting is available. Ethnicity information (Hispanic/non-Hispanic) has been more consistently tracked, with 95.8–100 percent of participants assigned ethnicity. Reporting of ethnicity was 100 percent complete for Phase 3 trials, where reporting was completed, and race information was available in at least 92.9 percent of trial participants in Phase 3 trials supported by NIH centers and institutes, suggesting the gaps in minority reporting originate predominantly in earlier-stage trials.

We have also assessed the quality of stratified results reporting in clinical trials registered at ClinicalTrials.gov with at least one site in the United States, with annual reporting by year of primary completion shown in Figures B-10a and B-10b (stratified by funder as reported by ClinicalTrials.gov, and the difference indicated between all phases and just Phase 3 [commonly registrational] trials). In both cases, trialists responded positively to new results reporting requirements, with NIH-funded trials reporting their results in greater than 50 percent of cases between 2008 and 2019, and industry funders reporting results in greater than 50 percent of cases between 2008 and 2018. Among Phase 3 trials, reporting of results was completed in at least 60 percent of industry-funded and NIH-funded trials between 2008 and 2019, with trials funded by other federal agencies and other entities achieving lower reporting compliance. Industry-funded trials account for more than two-thirds of Phase 3 trials (with much of the balance supported by the NIH, explaining the noisy data for other federal agency funding) and greater than 50 percent of all clinical trials (with NIH funding supporting 10–20 percent of all trials since 2008).

In July 2021, the FDA issued a Notice of Noncompliance to biotechnology company Accuitis, Inc., pursuant to 42 U.S.C. 282(j)(5)(C)(ii), requesting that results of its Phase 2 clinical trial of a treatment for acne rosacea be submitted to the ClinicalTrials.gov data bank (after first contacting the company about the issue in October 2020).[14] The company posted the results in August 2021.

---

[13] See https://www.census.gov/quickfacts/fact/table/US/PST045219.

[14] See https://www.fda.gov/media/151081/download.

APPENDIX B

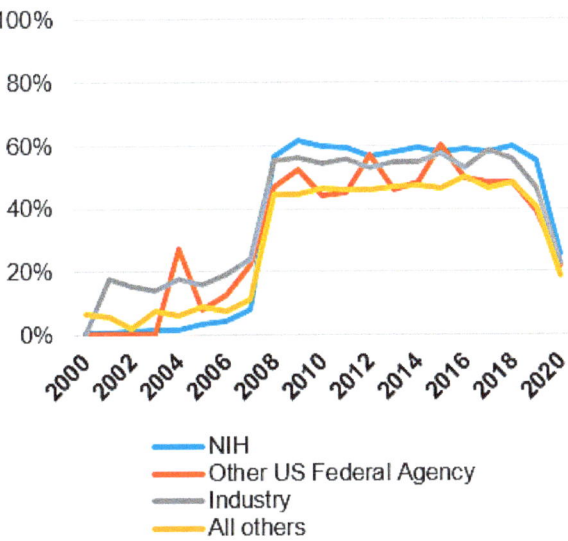

FIGURE B-10a Availability of results among all trials, by primary completion year, as of June 2021.

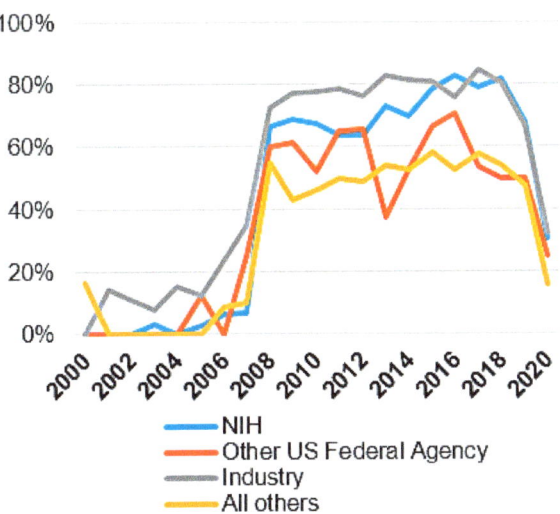

FIGURE B-10b Availability of results among Phase 3 trials, by primary completion year, as of June 2021.

Two other clinical trials (NCT03052816 and NCT01727336) have received an FDAAA 801 Notice status on ClinicalTrials.gov as of September 15, 2021, and their results have been submitted following FDA's notifications. This followed an April 2021 announcement by the FDA that it has sent more than 40 pre-notices of noncompliance to "encourage voluntary compliance with the ClinicalTrials.gov requirements"[15] and an August 2020 final guidance on civil money penalties related to noncompliance, which emphasized its regulatory attention will prioritize trials of products that "may pose a higher risk to human subjects or . . . products intended to address significant public health need," focusing on "responsible parties or submitters who have had a pattern of previous noncompliance with the requirements" and "applicable clinical trials for which noncompliance . . . exists in conjunction with noncompliance with other statutory and/or regulatory requirements" (HHS, 2020). It is possible that such regulatory enforcement will result in a higher rate of results reporting to ClinicalTrials.gov.

DISCUSSION

This analysis provided a brief summary of published research on diversity challenges of different types in clinical trial enrollment, and presented data on trial enrollment as reported by NIH institutes and centers for trials they sponsored, and by the FDA for trials that resulted in an approval.

The reporting of demographic representation in clinical trials has historically not been consistent and comprehensive. In this analysis, we find, for example, that race data have not been completely reported by NIH institutes and centers (however, ethnicity has been more reported with greater consistency). Moreover, the share of trials that report any results on ClinicalTrials.gov has increased sharply following the adoption of FDAAA in 2007. However, reporting compliance plateaued soon thereafter, with only 45–60 percent of all trials reporting results in most years. A higher share, up to 85 percent of industry-sponsored trials in Phase 3, have reported results in the database.

There may be several reasons for the lack of data on trial enrollment and outcomes, such as that the studies may not be required to submit results, are ongoing, have been completed but the deadline to post results has not passed, or the posting of results is pending certification or a request to extend the results submission deadline.[16] However, it may be important to better understand the reasons for the variance between reporting rates by trial sponsors, and to observe the effects of increased enforcement activity by the FDA in recent months.

---

[15] See https://www.fda.gov/news-events/press-announcements/fda-takes-action-failure-submit-required-clinical-trial-results-information-clinicaltrialsgov.

[16] See https://clinicaltrials.gov/ct2/about-site/results.

Trials Resulting in FDA Approval

The findings suggest several positive trends. First, in trials that resulted in FDA approval since 2014, the participation of females has consistently trended toward general parity (although notable differences between therapeutic areas still persist). Participation of females in non-oncology trials resulting in FDA approval has trended toward 60 percent of all participants, while their participation in oncology trials has plateaued just over 40 percent. Second, the participation of racial minorities in trials resulting in FDA approval has generally increased since 2014, with the number of white participants declining gradually to less than 75 percent in 2020. Finally, the participation of the elderly (over 65) has been on an increase, reaching nearly 40 percent in 2020. However, there are large differences by therapeutic area, with some of them (such as cardiovascular disease) showing above-average representation of the elderly, and others (such as endocrinology/metabolism/bone diseases, or gastroenterology) consistently below average.

NIH-Sponsored Trial Diversity

The findings show large differences in enrollment diversity over time and by NIH institute/center. For instance, while white participants have seen a slight increase across all trials (particularly in 2018, when they exceeded 60 percent for the first time, on average), their representation has been as low as 14.4 percent and as high as 81 percent in some cases. Representation of females has generally been consistent, ranging from 44.3 percent to 54.1 percent between 2013 and 2018, but has differed significantly by NIH institute/center. For example, at least 70 percent of participants in National Institute of Environmental Health Sciences–supported trials were females in all years for which data are reported, but the share of females was less than 40.5 percent in trials supported by the National Institute on Aging in 2017 and 2018 (this may be of particular concern in indications such as Alzheimer's disease and other dementias where female prevalence exceeds that of males).

Data on race and ethnic group representation in NIH-supported trials have not been consistently reported, with up to 22.4 percent of participants having unknown ethnicity in 2016. The participation of Hispanics has been most consistent, averaging less than 10 percent during most years for which data are available (except for 2016, when it reached 11.3 percent). This has also been a group that has been relatively the most underrepresented in clinical trials given the share of Hispanics in the total U.S. population. Black/African American participation has historically been between 10 percent and 25 percent (the National Cancer Institute has had the lowest success in this category in 2016 and 2017, when it dropped under 5 percent). Notably, the National Institute of Allergy and Infectious Diseases has reported shares of greater than 30 percent (except for 28.1 percent in 2018). Representation of Asian participants has been somewhat less consistent, with some years showing spikes by some NIH institutes/centers (mostly driven by a 2017 increase in Asian recruitment by trials supported by the National Cancer Institute). However, it appears that the trend has been otherwise a decreased one, reaching 7.8 percent in 2018, from 16.3 percent in 2013.

CONCLUSIONS

Some promising trends indicate more diverse trial enrollment in trials resulting in FDA approval as well as different diversity profiles of trials sponsored by NIH institutes/centers. The representativeness of trials resulting in an FDA approval has been gradually improving in several dimensions since 2014. The diversity profile of trial participants in NIH-sponsored trials has not followed a similar trend, however. As shown in this analysis, significant differences between different NIH institutes and centers have emerged (aside from fluctuations over time). These differences may be explained by numerous factors, such as variable disease prevalence in select populations. A study of the particular drivers of enrollment differences is warranted, adjusting for unique demographic profiles of multicountry trials (as population demographics in most regions of the world are very different from the U.S. demographic profile).

This study has highlighted that reporting gaps exist for clinical trials supported by public funding as well as for trials that have been approved. For example, there is no broadly available tool to analyze diversity of all trials resulting in FDA approval via a publicly available database (requiring manual collection and analysis of FDA Snapshot data, which makes any ongoing analysis more challenging). Such an approach is more prone to errors and is less flexible with different reporting and analytic objectives. Moreover, the demographics of NIH-sponsored trials are reported in an aggregate form by NIH institutes and centers, preventing an analysis at a more granular level. Some institutes and centers, moreover, only report extramural research or do not make explicit distinctions between intramural and extramural research, and provide limited contextual information that would help explain large differences in total or subgroup enrollment data in some years. More detailed reporting of trial participation at the trial level in a publicly accessible database would allow for a more informative analysis of key trends in NIH-funded clinical trials, further contributing to transparency and accountability. The NIH has recently expanded its reporting to stratify by disease area, but the lack of historical data has prevented a longitudinal analysis of trends to date.

The FDA's actions over 2021 suggest that compliance with reporting requirements will become a greater priority, which in turn may provide better data on demographic profiles of individual trials. It is yet to be seen whether reporting rates increase by commercial and/or other trialists as a result of increased enforcement activity, particularly for late-stage trials. Not only is the reporting of results and demographic characteristics helpful in our understanding of treatment effects and the generalizability of presented findings, but it also makes scientific progress more efficient by enabling researchers to review negative findings, compare results for subgroups across trials, and conduct analyses of unpublished trial results. Ultimately, greater transparency of reporting on enrollment and outcomes will benefit both clinical investigators and patients by increasing the information value of trials that do not result in FDA approval, as well as trials that reach regulatory review. A better understanding of trial enrollment and results reported for specific demographic populations is likely to accelerate scientific progress and ultimately lead to more effective treatment options for all patients.

Acknowledgments

The author is extremely grateful for research assistance provided by Austen Applegate and Alex Helman, both staff of the Committee on Women in Science, Engineering, and Medicine of the National Academies of Sciences, Engineering, and Medicine, and to the members of the committee and others who have provided support and constructive feedback.

REFERENCES

Anderson, G. D. 2005. Sex and racial differences in pharmacological response: Where is the evidence? Pharmacogenetics, pharmacokinetics, and pharmacodynamics. *Journal on Women's Health* 14(1):19–29. doi:10.1089/jwh.2005.14.19.

Artiga, S., J. Kates, J. Michaud, and L. Hill. 2021. Racial diversity within COVID-19 vaccine clinical trials: Key questions and answers. *KFF Health Policy*. https://www.kff.org/racial-equity-and-health-policy/issue-brief/racial-diversity-within-covid-19-vaccine-clinical-trials-key-questions-and-answers/.

Califf, R. M., D. A. Zarin, J. M. Kramer, R. E. Sherman, l. H. Aberle, and A. Tasneem. 2012. Characteristics of clinical trials registered in ClinicalTrials.gov, 2007–2010. *JAMA* 307(17):1838–1847. doi:10.1001/jama.2012.3424.

Carcel, C., and M. Reeves. 2021. Under-enrollment of women in stroke clinical trials. *Stroke* 52(2):452–457. https://doi.org/10.1161/STROKEAHA.120.033227.

Clark, L. T., L. Watkins, I. L. Piña, M. Elmer, O. Akinboboye, M. Gorham, B. Jamerson, C. McCullough, C. Pierre, A. B. Polis, G. Puckrein, and J. M. Regnante. 2019. Increasing diversity in clinical trials: Overcoming critical barriers. *Current Problems in Cardiology* 44(5):148–172. https://doi.org/10.1016/j.cpcardiol.2018.11.002.

Eshera, N., H. Itana, L. Zhang, G. Soon, and E. O. Fadiran. 2015. Demographics of clinical trials participants in pivotal clinical trials for new molecular entity drugs and biologics approved by FDA from 2010 to 2012. *American Journal of Therapeutics* 22(6):435–455. doi:10.1097/mjt.0000000000000177.

FDA (U.S. Food and Drug Administration). 1993. Study and Evaluation of Gender Differences in the Clinical Evaluation of Drugs: Guidance for Industry. https://www.fda.gov/regulatory-information/search-fda-guidance-documents/study-and-evaluation-gender-differences-clinical-evaluation-drugs.

FDA. 2011. Dialogues on Diversifying Clinical Trials. September 22–23, 2011, Washington, D.C.

FDA. 2020. Enhancing the Diversity of Clinical Trial Populations—Eligibility Criteria, Enrollment Practices, and Trial Designs Guidance for Industry. https://www.fda.gov/media/127712/download.

Flores, L. E., W. R. Frontera, M. P. Andrasik, C. del Rio, A. Mondríguez-González, S. A. Price, E. M. Krantz, S. A. Pergam, and J. K. Silver. 2021. Assessment of the inclusion of racial/ethnic minority, female, and older individuals in vaccine clinical trials. *JAMA Network Open* 4(2):e2037640–e2037640. doi:10.1001/jamanetworkopen.2020.37640.

HHS (U.S. Department of Health and Human Services). 2020. Civil Money Penalties Relating to the ClinicalTrials.gov Data Bank. Food and Drug Administration Guidance Document, 85 FR 50028: 50028–50030.

Lewis, J. H., M. L. Kilgore, D. P. Goldman, E. L. Trimble, R. Kaplan, M. J. Montello, M. G. Housman, and J. J. Escarce. 2003. Participation of patients 65 years of age or older in cancer clinical trials. *Journal of Clinical Oncology* 21(7):1383–1389. doi:10.1200/jco.2003.08.010.

Ludmir, E. B., W. Mainwaring, T. A. Lin, A. B. Miller, A. Jethanandani, A. F. Espinoza, J. J. Mandel, S. H. Lin, B. D. Smith, G. L. Smith, N. A. VanderWalde, B. D. Minsky, A. C. Koong, T. E. Stinchcombe, R. Jagsi, D. R. Gomez, C. R. Thomas Jr., and C. D. Fuller. 2019. Factors associated with age disparities among cancer clinical trial participants. *JAMA Oncology* 5(12):1769–1773.

Mak, W. W. S., R. W. Law, J. Alvidrez, and E. J. Pérez-Stable. 2007. Gender and ethnic diversity in NIMH-funded clinical trials: Review of a decade of published research. *Administration and Policy in Mental Health and Mental Health Services Research* 34(6):497–503. doi:10.1007/s10488-007-0133-z.

Malzbender, K., L. Lavin-Mena, L. Hughesm N. Bose, D. Goldman, and D. Patel. 2020. *Key Barriers to Clinical Trials for Alzheimer's Disease*. Los Angeles: University of Southern California Leonard D. Schaeffer Center for Health Policy & Economics. https://healthpolicy.usc.edu/wp-content/uploads/2020/08/Key-Barriers-to-Clinical-Trials-for-Alzheimer%E2%80%99s-Disease_FINAL.pdf.

Murthy, V. H., H. M. Krumholz, and C. P. Gross. 2004. Participation in cancer clinical trials: Race-, sex-, and age-based disparities. *JAMA* 291(22):2720–2726. doi:10.1001/jama.291.22.2720.

Sedrak, M. S., R. A. Freedman, H. J. Cohen, H. B. Muss, A. Jatoi, H. D. Klepin, T. M. Wildes, J. G. Le-Rademacher, G. G. Kimmick, W. P. Tew, K. George, S. Padam, J. Liu, A. R. Wong, A. Lynch, B. Djulbegovic, S. G. Mohile, W. Dale; Cancer and Aging Research Group (CARG). 2021. Older adult participation in cancer clinical trials: A systematic review of barriers and interventions. *CA: A Cancer Journal for Clinicians* 71(1):78–92. https://doi.org/10.3322/caac.21638.

Strong, B., V. Howard, and M. Reeves. 2020. Sex differences in enrollment among contemporary acute stroke trials: Differences by trial type. Paper presented at the ESO-WSO 2020 Virtual Conference, November 7–9, 2020. *International Journal of Stroke*. https://journals.sagepub.com/doi/full/10.1177/1747493020963387.

Woodall, A., C. Morgan, C. Sloan, and L. Howard. 2010. Barriers to participation in mental health research: Are there specific gender, ethnicity and age related barriers? *BMC Psychiatry* 10(1):103. doi:10.1186/1471-244X-10-103.

APPENDIX B

SUPPLEMENTARY MATERIAL

The Supplementary Table, below, shows the demographic representation by category in Phase 3 trials supported by NIH centers and institutes. Note that several demographic characteristics are not stable over time, driven by several large trials run by some research centers and institutes, or incomplete reporting by all centers and institutes over time. Data for Phase 3 trials have a smaller share of participants with missing race or ethnicity information than all trials supported by NIH centers and institutes.

SUPPLEMENTARY TABLE Demographics of Participants in Phase 3 Trials Supported by NIH Centers and Institutes

	2013 (%)	2014 (%)	2016 (%)	2017 (%)	2018 (%)
Female	64.2	65.9	53.1	59.7	61.5
American Indian	5.2	6.5	0.3	0.4	0.7
Asian	33.6*	42.0*	5.2	4.4	5.4
Black/African American	20.5	20.9	67.5	14.8	17.0
Native Hawaiian/Pacific Islander	0.1	0.1	0.0	0.1	0.2
White	35.5	25.6	22.6	72.0	64.3
More than 1 race	0.5	0.4	0.3	0.6	3.7
Unknown race	0.5	0.4	0.3	0.6	3.7
Hispanic	12.4	12.0	11.9	58.5**	40.0**
Non-Hispanic	86.9	86.2	86.9	30.4	56.7
Unknown ethnicity	0.7	1.8	1.2	11.1	3.3
Sum of all races	95.9	95.9	96.3	92.9%	94.9
Sum of all ethnicities	100.0	100.0	100.0	100.0	100.0%

* Data on Asian enrollment in 2013 and 2014 were affected by large Asian representation in trials run by the National Institute of Child Health and Human Development (for which enrollment data are not available for 2016–2018).

** The large increase in Hispanic representation was driven by significant increases in reported Hispanic participants by the National Heart, Lung, and Blood Institute (from about 41,200 in 2016 to more than 317,000 in 2017 and nearly 118,500 in 2018).

Appendix C

Improving Representativeness in Clinical Trials and Research: Facilitators to Recruitment and Retention of Underrepresented Groups

*Franchesca Arias, Ph.D. (1), (2), (3); *Nicole Rogus-Pulia, Ph.D., C.C.C.-S.L.P. (4), (5); and Amy J. H. Kind, M.D., Ph.D. (4), (5), (6)

Affiliations:

(1) Aging Brain Center, Hinda and Arthur Marcus Institute for Aging Research at the Hebrew SeniorLife, Boston, Massachusetts
(2) Department of Neurology, Beth Israel Deaconess Medical Center, Boston, Massachusetts
(3) Harvard Medical School, Boston, Massachusetts
(4) Department of Medicine, Division of Geriatrics and Gerontology, University of Wisconsin-Madison, School of Medicine and Public Health, Madison, Wisconsin
(5) Madison VA Geriatrics Research Education and Clinical Center, Middleton VA Hospital, Madison, Wisconsin
(6) Center for Health Disparities Research, University of Wisconsin–Madison, Madison, Wisconsin

*Contributed equally as first authors.

ABSTRACT

Objective: Clinical trials are essential for determining safety and efficacy of health-related interventions as well as informing future research and funding priorities in the United States. However, recruitment and retention challenges result in underrepresentation of diverse groups in clinical trials, which limits understanding of disease mechanisms as well as generalizability of findings. The purpose of this study was to elucidate facilitators to recruitment and retention strategies of underrepresented groups, based on a representative sample of published clinical trials with successful inclusion.

Method: A mixed-methods approach was employed to accomplish these aims. Research teams with experience recruiting underrepresented groups were invited to participate in individual comprehensive interviews. Twenty interviews were completed that focused on understanding facilitators to recruitment and retention into clinical trials. To identify studies appropriate for in-depth qualitative interviews, we first conducted a systematic review of published clinical trials available on PubMed between 2001 and 2021 across the top six diseases leading to mortality in the United States (heart disease, cancer, chronic lower respiratory disease, stroke, Alzheimer's disease, and diabetes). From these trials, we randomly selected 162 trials stratified by disease and geographic location. We then benchmarked the number of study participants by race, ethnicity, and sex against the local (single site) or national (multisite) data as reported by the American Community Survey Demographic and Housing Estimates years 2010 to 2019. Study teams for those trials that met diversity criteria (50 percent or greater recruitment in at least one category) were invited to participate in the interviews.

Results: Of the 162 randomly selected trials, 142 met diversity criteria following benchmarking. Incomplete reporting of sample characteristics was observed in the majority of studies; however, 96 percent of trials reported the sex of their study participants. Of the trials that achieved success in recruiting representative samples, less than 33 percent reported information about ethnicity and less than 66 percent of trials included a robust breakdown of the racial representation. Of the 142 study teams invited, 20 participated in the interviews. Results from qualitative analysis interview transcripts revealed eight main themes with associated subthemes: (1) starting with intention and agency to achieve representativeness; (2) establishing a foundation of trust with participants and community; (3) anticipating and removing barriers to study participation; (4) adopting a flexible approach to recruitment and data collection; (5) building a robust network by identifying all relevant stakeholders; (6) navigating scientific, professional peer, and social expectations; (7) optimizing study team to ensure alignment with research goals; and (8) attaining resources and support to achieve representativeness.

Discussion: While issues of representativeness in research have been at the forefront of science in the past 10 years, additional efforts are necessary to systematically assess and comprehensively report social and cultural characteristics of cohorts in peer-reviewed publications of clinical trials. While intentionality drives current scientific efforts to understand how diseases affect persons from diverse groups, this work remains underfunded and undervalued. A call to action that involves providing resources, expanding the definition of stakeholders, integrating community-based stakeholders as equitable partners, and involving national funding organizations, academic institutions, and the scientific community is necessary to meaningfully advance work in this area.

Keywords: recruitment, retention, minority, representation, clinical trial

INTRODUCTION

Clinical trials provide the most robust evidence to document the efficacy and safety of pharmacological and non-pharmacological interventions. Moreover, clinical trial evidence is often used to identify areas for future research as well as to guide funding priorities and allocation of resources. However, recruitment challenges often hinder the utility and generalizability of clinical trials. Recent data from the National Institutes of Health (NIH) indicate that less than 20 percent of clinical trials in the United States meet their recruitment targets, and up to 80 percent of these studies require extensions as a result of low enrollment (Clinical Trials Arena, 2012). Recruitment challenges are even more pronounced when considering representation of diverse groups.

Representation in clinical trials is particularly important in the context of the changing U.S. demographics. By 2045, it is anticipated that nearly half of the U.S. population will self-identify as ethnoracially diverse (Census, 2018). Persons who self-identify as Black/African American, American Indians/Alaska Native, Asian/Asian American, Native Hawaiian, or Hispanics/Latino(a) are more likely to be poor and underinsured. Moreover, persons from these and other historically underrepresented groups experience increased disease burden from common conditions such as heart disease, diabetes, asthma, obstructive pulmonary disease, stroke, obesity, and liver disease (Carratala and Maxwell, 2020) For example, persons who self-identify as lesbian, gay, bisexual, transgender, or queer (LGBTQ+) have unique health experiences and are disproportionately affected by mental health conditions and sexually transmitted diseases (SAMHSA, 2012). There is substantial underrepresentation of diverse groups in clinical research. Data published by the Food and Drug Administration (FDA) suggest that participants enrolled in clinical trials for pharmacological interventions overwhelmingly self-identified as non-Hispanic white (81 percent), with 4 percent Black/African American and 12 percent Asian/Asian American. Ethnicity is inconsistently reported. When available, Hispanic/Latina(o) represent about 11 percent of patients enrolled in pharmacological trials (Duma et al., 2018; FDA, 2020; Frew et al., 2014; Gong et al., 2019; Khan et al., 2020; McCarthy-Keith et al., 2010; Rosende-Roca et al., 2021; Vitale et al., 2016). These numbers do not reflect the current U.S. population, in which 14 percent of persons in the United States identify as Black/African American, 7 percent as Asian/Asian American, 2 percent as American Indian, and 18 percent as Hispanic/Latino(a) (Census, 2019). Lack of adequate representation threatens the integrity of science. Interventions evaluated on a subset of the population and under circumscribed settings may not realistically generalize to other groups and settings (Haidich and Ioannidis, 2001). For certain medical conditions (e.g., asthma, heart failure, cancer), drug response profiles may differ based on ethnoracial factors (Jamerson and DeQuattro, 1996; Tay et al., 2020); however, there is not currently enough representation of low-income and non-white persons in drug trials to determine whether social or biological factors are associated with differential responses to drugs (Odierna and Bero, 2009). Finally, lack of representation and limited reporting on the social and contextual factors influencing disease trajectories may interfere with replicability of findings (Glasgow et al., 2018) and our ability to identify mechanisms underlying diseases (Ix et al., 2008).

The NIH has implemented initiatives designed to foster the inclusion of underrepresented groups in NIH-supported clinical research trials (NIH, 2001). Similarly, the FDA implemented reporting requirements and issued a recommendation for sponsors of clinical trials to increase enrollment of underrepresented populations (FDA, 2014). Nevertheless, underrepresentation of diverse groups in clinical trials persists (Nazha et al., 2019). Engaging in strategies that will ultimately increase recruitment and retention of underrepresented groups in clinical trials will result in diverse samples that more appropriately reflect the population who will ultimately utilize, and who stands to benefit from, the intervention. Importantly, increased representativeness in clinical trials assures the efficacy and safety of treatments in these diverse subgroups.

The purpose of this study was to characterize current efforts on representativeness in clinical research and to systematically assess effective recruitment and retention strategies. Finally, we hope to recommend strategies that can be used by scientists to diversify clinical trial participant populations.

METHODS

We developed a novel Systematic Randomized Qualitative Assessment (SRQA) methodology that incorporated both quantitative and qualitative techniques to answer our question in a balanced and inclusive manner. Since our goal was to elucidate facilitators for recruitment and retention strategies of underrepresented groups, we first needed to identify studies with successful recruitment. Thus, we performed a systematic search of clinical trials published in PubMed between 2001 and 2021. We aimed to identify U.S.-based clinical trials that successfully recruited historically underrepresented groups as assessed by objective population-representative benchmarking criteria (described below). Figure C-1 offers an overview of this process according to the 2012 PRISMA guideline for systematic reviews (Moher et al., 2012).

Inclusion and Exclusion Criteria

The following inclusion criteria were applied to the studies identified from the systematic review: (1) manuscripts written in English, (2) recruitment conducted only in the United States, (3) publication in a peer-reviewed journal, (4) inclusion of adult participants (age 18 years and over), and (5) recruitment completed by time of publication. Secondary analyses were included if prospective data were collected, and the new data were available to complete benchmarking, or if original data were available for benchmarking.

First, we prioritized our search in clinically relevant areas. As such, we focused our search on clinical trials addressing the top six causes of mortality in the United States according to the Centers for Disease Control and Prevention (CDC): heart disease, cancer, chronic lower respiratory disease, stroke, Alzheimer's disease, and diabetes. Unintentional injuries and intentional self-harm were excluded to maintain a focus on medical conditions (CDC, 2020). Our exclusion criteria were (1) non-interventional studies, (2) dissertations, (3) non-human studies, (4) case studies, and (5) meta-analyses (see Figure C-1).

Given that representation of diverse groups differs across U.S. regions (Frey, 2019), our methodology sought to ensure that all areas of the United States were represented in the search. To this end, we stratified our search by using geographic filters in PubMed based on the nine U.S. Census Divisions: New England, Middle Atlantic, East North Central, West North Central, South Atlantic, East South Central, West South Central, Mountain, and Pacific (Lewison, 1997). These filters were based on the address of the corresponding author for the published trial and were added to search terms for each disease category. Given that PubMed is the only available search engine that allows for this type of custom geographic filter specific to U.S.-based trials, other search engines were not included (see Appendix C-1 for search terms that were applied).

Once the study was selected, we performed full text review to confirm the study recruitment location, and any discrepancies were resolved. To avoid penalizing research studies located in regions with low representation of target groups, studies were benchmarked to data from the American Community Survey Demographic and Housing Estimates years 2010 to 2019 collected by the U.S. Census Bureau using the region of recruitment. County-level data were used for single-site studies. Multisite studies conducted within the same state were benchmarked using state-level data. Studies that included sites dispersed across multiple states were benchmarked using national-level data.

Our goal in this stage was to identify a group of trials, stratified by disease condition and geographic division, which met our prespecified definition of diverse recruitment. In total, three trials from each of the nine regions for each of the six disease categories were to be identified (3 x 9 x 6 = 162). To accomplish this, we first stratified the trials into six CDC disease categories and then subdivided the strata into the nine U.S. Census divisions. Next, we randomly selected studies from each division by disease cell and applied our inclusion and exclusion criteria until an eligible trial was identified. We repeated this process until we reached the target number of three eligible trials per disease condition by geographic division (n = 162) (see Table C-1).

Data Extraction

Once the final set of 162 eligible trials was identified, data extraction was carried out independently by four reviewers using an extraction form specifically designed for the purpose (see Appendix C-2). The form was pilot tested for feasibility and reliability on five sample trials prior to use. Extraction was completed in duplicate for each trial, and any discrepancies between reviewers were resolved through consensus. Data were extracted on clinical trial study design, study setting, study inclusion and exclusion criteria, and interventions. In addition, information about the characteristics of the included study sample (e.g., age, race, ethnicity, sex/gender, gender identity, sexual orientation, disability status) was extracted, with a special focus on the demographics for underrepresented groups. Underrepresented groups were identified according to the fundamental causes of disparities listed in the National Institute on Aging (NIA) Health Disparities Framework (Hill et al., 2015).

Variable Definitions

Representativeness of the cohort was reported using a multilevel approach.

Level 1: NIH-Mandated Diversity Variables

These diversity variables were defined as characteristics currently required for reporting by the NIH in the targeted enrollment tables that are mandatory for every clinical trial since March 1994 (Taylor, 2008). These variables include the number and proportion of participants recruited by sex, ethnicity, and race with data presented according to well-established classifications (Riley et al., 2018). Detailed information about ethnic/racial groups was abstracted when available, as follows: (1) for Hispanic/Latino(a) participants, we reported whether a cohort was from the Caribbean, Central America, North America, or South America; (2) for Asian and Pacific Islander participants, we reported whether the cohort was predominantly from Eastern Asia, South Central Asia, Southeastern Asia, or Western Asia; (3) for American Indian or Alaska Native participants, we reported whether the cohort was predominantly from the Great Plains Area, Alaska Area, Albuquerque Area, Bemidji Area, Billings Area, California Area, Nashville Area, Navajo Area, Oklahoma Area, Phoenix Area, Portland Area, or the Tucson Area; (4) for African American or Black participants, we reported whether the cohort was predominantly from African origin, Caribbean, and Other origin.

Level 2: Intersectionality

Information about participants' intersectionality, or the interconnectedness among social groups and how these intersections confer unique privilege or vulnerabilities, was collected if these characteristics were described in the published manuscript (e.g., multirace; LGBTQ+ participant from racially underrepresented groups) (Crenshaw, 2017). Intersectionality of study participants (sex + race + ethnicity) was recorded in a binary form (characterized or not characterized).

Level 3: Other Underrepresented Groups

Recruitment of individuals from other underrepresented groups (80 years and older, disability, gender identity, and sexual orientation) was also recorded in binary form (characterized or not characterized for each diversity category).

Outcomes/Benchmarking

The primary outcome of interest was the diversity achieved in each study cohort as compared to census data for the specific region, time frame, and specific diversity variable of interest. The goal of our rubric was to assess representativeness for each trial, taking into consideration the characteristics of the population available to those researchers in their communities. As such, to avoid penalizing research studies located in regions with low representation of target groups, studies were benchmarked to data from the American Community Survey Demographic and Housing Estimates years 2010 to 2019 (U.S. Census Bureau data). For single-site studies, county-level data were used. For multisite studies, state-level data or national-level data were used, depending on whether sites were within the same state or dispersed across the United States. "Successful" studies were those that recruited 50 percent or greater of the proportional target population in their region (i.e., 50 percent of the county-level base rate, the state-level base rate, or the national-level base rate) for the targeted demographic characteristic (sex, ethnicity, or race). This process was independently assessed by two reviewers.

Qualitative Study: In-Depth Interviews

The next step was to conduct in-depth qualitative interviews to provide a comprehensive assessment of their recruitment strategies, with particular interest in practical examples of successful and innovative strategies. To this end, the corresponding author on each benchmarked trial with "successful" recruitment of diverse groups was invited to participate (or to designate an appropriate recruitment staff member to participate) in these interviews. We contacted 142 corresponding authors for studies that met the inclusion criteria via email. Interview requests were submitted in batches of 30. All eligible trials were contacted at least twice. Of these, 40 investigators responded, 5 of whom declined participation. The most common reasons for declining participation included lack of involvement in study recruitment or retention and lack of time for the interviews. Scientists from 20 studies ultimately agreed to participate in the interviews (see Figure C-1).

Our interview guide (see Appendix C-3), which focused on facilitators to recruitment and retention of diverse groups, was pilot tested and iteratively refined during the first four interviews. Interviews were conducted with individuals using open-ended questions, such as "Tell us about the strategies you implemented to enhance recruitment of underrepresented populations in your study." Specific probes asked such questions as "Did you engage stakeholders in your study? If yes, at what point in the process were they involved?" The *Model Framework of Multilevel Factors Affecting Decision to Participate in Clinical Trial*, which was proposed by Ford et al. (2013), guided the development of our clinical interview and general approach to the qualitative portion of this project.

To ensure rigorous data collection, all interviews were conducted by video, audio-recorded, and transcribed verbatim with all identifying names or other personal health information removed. Interviews lasted 45 minutes on average, and we conducted interviews until adequate saturation was reached for thematic content analysis (no new concepts detected for at least three interviews).

Qualitative data were iteratively coded, sorted, and compared using thematic analysis by two raters (Boyatzis, 1998). An initial codebook was developed during piloting, and two raters separately identified tentative themes and subthemes using data from the pilot interviews by conducting line-by-line coding using NVivo 11 software (QSR International). After reconciling differences in the inductive codes, we updated the coding scheme. Knowledge gained from each interview was incorporated into subsequent sessions to refine questions and explore salient themes. During coding, inter-rater reliability was assessed and any discrepancies were resolved through consensus. To ensure the rigor of the analytic approach, we implemented several processes, including peer debriefing, independent and collaborative coding, refinement of themes by examining supporting and contradictory cases, and documentation of a decisional audit trail (Frey, 2019).

This study was reviewed by the institutional review boards (IRB) of the National Academies of Sciences, Engineering, and Medicine, the University of Wisconsin–Madison, and Beth Israel Deaconess Medical Center/Harvard Medical School, and was determined to be exempt from IRB review.

RESULTS

Quantitative Categorization of the Diversity of Study Cohorts

A total of 131,028 clinical trials were identified using the search strategy in Appendix C-1 across all six disease categories and nine geographic divisions. Numbers within each category of disease and geographic region are displayed in Table C-1.

A total of 1,279 clinical trials were randomly selected, stratified by disease category and geographic region. Careful review of abstracts and, when available, trial information on ClinicalTrials.gov, was performed applying inclusion and exclusion criteria to identify 162 eligible trials (three trials for each of the six CDC disease categories across the nine U.S. Census regions). More than 1,100 trials were excluded (see Figure C-1 for exclusion reasons).

TABLE C-1 Number of Trials in Each Disease Category by U.S. Census Geographic Regions and Divisions

U.S. Census Regions	U.S. Census Divisions	Heart Disease	Cancer	Chronic Lower Respiratory Disease	Stroke	Alzheimer's Disease	Diabetes
1 - Northeast	1. New England	3,836	1,350	434	337	141	5,429
	2. Middle Atlantic	5,410	2,103	580	422	176	6,349
2 - Midwest	3. East North Central	4,266	1,474	429	347	151	5,888
	4. West North Central	2,848	838	288	221	85	3,272
3 - South	5. South Atlantic	12,161	4,311	1,643	972	292	19,306
	6. East South Central	2,176	587	180	161	27	3,056
	7. West South Central	4,144	1,851	400	346	104	6,008
4 - West	8. Mountain	2,277	1,009	405	190	126	4,327
	9. Pacific	5,425	2,264	688	451	243	9,224

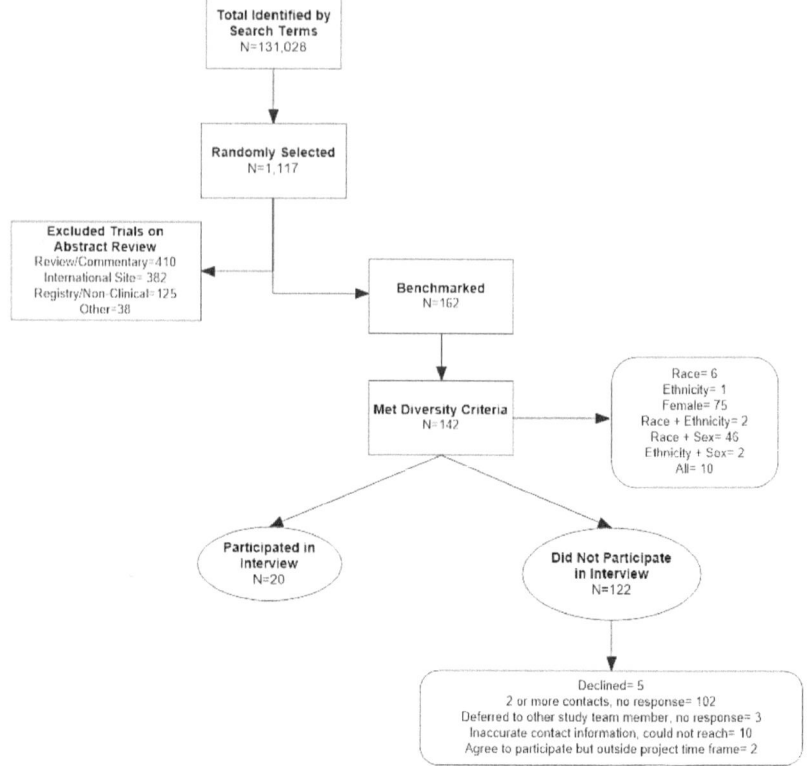

FIGURE C-1 Flow chart illustrating process for identification of trials.

Met Diversity Criteria

Study teams met diversity criteria if they recruited 50 percent or greater in at least one category: race, ethnicity, and sex relative to county (single site) or state/national (multisite) data as reported by the American Community Survey Demographic and Housing Estimates years 2010 to 2019.

Did Not Participate in Interview

Researchers that met diversity criteria were contacted sequentially in batches of 30. All eligible researchers were contacted twice within a 3-week period.

Benchmarked Trials

All 162 eligible trials were benchmarked in each of the NIH-mandated diversity categories (sex, race, and ethnicity) against local county-level U.S. Census data (for single-site study) and state- and national-level data (see Appendix C-4). Of the 162 eligible trials, 142 trials (88 percent) met the preestablished diversity criteria of 50 percent or greater recruitment in at least one of the three NIH-mandated categories (sex, race, and ethnicity). Of those, 53 percent met criteria for recruiting female participants only, and about 1 percent did not describe their sample in terms of sex; 64 percent of the trials that met the preestablished diversity criteria did not report details on the ethnicity of their participants; and 3 percent of trials met criteria for one of the four racial groups recorded by the U.S. Census (African American or Black [n = 4], American Indian or Alaska Native [n = 1], Native Hawaiian or Pacific Islander [n = 0], or Asian [n = 0]). One trial met criteria for recruiting multiple races (n = 1). About 46 percent of trials did not report information on the racial breakdown of their sample, and many reported race as "White" and "Other than White." Thirty-two percent of trials met criteria for sex and race; 1 percent met criteria for sex and ethnicity; 1 percent met criteria for race and ethnicity; and 7 percent met criteria for all three categories.

Regarding intersectionality, only 5 percent of the trials reported whether participants in their sample self-identified as belonging to more than one underrepresented group. Similarly, only 7 percent of studies reported information on the disability status of their participants. Self-reported gender identity and/or sexual orientation were not included in any of the trials reviewed.

Qualitative Assessments of Trials that Succeeded in Recruiting a Diverse Cohort

A total of 20 study teams and 22 participants were included in the qualitative portion of the study (18 single and 2 two-person interviews). The average age of participants was 49.8 (13.9) [M(SD)] with a range of 27 to 73. Overall, 74 percent self-identified as female, and no participant self-identified as nonbinary. Participants self-identified as Hispanic/Latina(o) (11 percent), non-Hispanic white (53 percent), Black/African American (11 percent), and Asian (26 percent). Native American/Alaska Native or Native Hawaiian/Pacific Islander scientists were not represented in this study.

In the thematic analysis, eight major themes emerged that were broadly related to representativeness in clinical research: (1) starting with intention and agency to achieve representativeness; (2) establishing a foundation of trust with participants and community; (3) anticipating and removing barriers to study participation; (4) adopting a flexible approach to recruitment and data collection; (5) building a robust network by identifying all relevant stakeholders; (6) navigating scientific, professional peer, and social expectations; (7) optimizing study team to ensure alignment with research goals; and (8) attaining resources and support to achieve representativeness. Major themes and subthemes are listed and described below. Additional details and quotes are provided in Tables C-2a to C-2h.

Starting with Intention and Agency to Achieve Representativeness

Participants described the work with diverse communities as resource, time, and labor intensive. For example, "It's a lot of work and a lot of time and it takes years.... We've been working with the same community partners now for 12, 13 years. They see us all the time." They explained that the intention must be based in the reality that a multistage process is required to achieve representativeness. That is, contact with community members begins long before recruitment and extends long after research support ends, and resources ought to continue to effect changes in communites long after the study ends. Participants emphasized that collaboration with community members specific to recruitment and retention strategies occurs across different stages of the study. For example, "I think some of the principles that are laid out for stakeholder engagement basically involve them in the design of the study, the conception of the study, what questions you're asking, as well as in how you're doing, the recruitment, who you're recruiting, what your materials are, and then what the study involves, like kind of soup to nuts kind of thing. And so I try to do that as much as I can."

Greater than 80 percent of the participants reported that being intentional about having representation of historically underrepresented groups was instrumental to their success. Intrinsic motivation stems from a personal commitment to promote equity and erradicate health disparities, ethical and professional values fueling the desire to elucidate the biological and contextual mechanisms driving the effectiveness of an intervention, and a genuine scientific curiosity to understand disease in different patient groups. For example, "We don't know how they respond to different interventions. We just don't know what the differences are, we don't have nearly enough data," and "I need to tell you about a comment from one of the Black leaders ... when we were discussing the ... drug ... that has only seven percent African-American and then one Native American in the entire study.... This is disrespecting the Black body in the same way as slavery. You're not respecting people. You're the ones that sit at these tables where we are not, how dare you all put out a drug that everyone can't use as if we're not dying from this disease."

Frequent subthemes were the perception that clinical trials provided access to innovative treatments. For example, "It's absolutely important in terms of behavioral interventions and how you implement [with] certain people or not if you don't have access to the things that people of high social economic status take for granted. If you don't have that kind of access, then you're not going to be as able to implement any intervention, especially behavioral ones that require changes in lifestyle, taking time out of your day and stuff like that." Under these themes, the view that research is an endeavor largely funded by taxpayers that should benefit taxpayers across all socioeconomic levels was also evident. For example, when I go to the talks, I say, this is your ... these are your tax dollars at work. This is your money. You need to benefit from this too. You and your friends and family, let's get everybody on board." Extrinsic motivation came from external factors, including requirements by funding agencies, parameters imposed by the environment (e.g., need to recruit from a given state or setting), and factors driven by the characteristics of the diseases, such as base rates.

Establishing a Foundation of Trust with Participants and the Community at Large
The idea of building and maintaining trust with both study participants and their larger communities was reported by 100 percent of the participants. History of abuse by researchers, experiences with other research groups that approached underrepresented communities for the purposes of a study and did not remain engaged, and beliefs that research is not beneficial to the community were cited as barriers to establishing trust with persons from diverse communities. For 85 percent of the participants, the development of trust requires a long-term commitment by the principal investigators, study team, and local institutions that benefit from this research. Respondents described the necessity of building trust over time through consistent engagement in the community, developing meaningful relationships with study participants, and giving to the community without the expectation of anything in return. Many participants emphasized that while trust has to be built over time, trust can be broken with individuals and communities in an instant. For example, "There's such trust building, that . . . takes awhile. And if one person drops and doesn't keep the trust, then I'm not going to be able to most likely get back that location again." An approach to community partnership that is truly equitable and not hierarchical in nature (15 percent) was suggested as a way to mitigate distrust in these communities and to most effectively leverage resources for truly meaningful and translatable work. For example:

> I think that is the goal to get to full equity with the community partner, writing the grants and getting the money and sharing everything from the ground up to the study. I think we're still unequal with academic partners. So doing a grant writing, getting the funding and working with community partners and giving them funding from the grant. So I think there's still this hierarchy. Unfortunately, we're trying to break those down. We're trying to get to parity as much as possible. And that's just going to take time and it's going to take investment.

According to 20 percent of the participants, in addition to facililtating recruitment, establishing relationships with community leaders provided opportunities to understand the needs of the community in order to build trust over time.

Anticipating and Removing Barriers to Study Participation
Included here are aspects of clinical trial participation that may influence accessibility to research for members of underrepresented communities. High-priority subthemes focused on barriers to participation and removal of these barriers. Systemic barriers included complex consent language, lengthy research visits, research activities that place undue burden on participants (e.g., requiring them to miss work), or issues related to physical access (e.g., driving many hours to participate in person) that are unique to the research enterprise. Other barriers were sociocultural and revolved around the goodness of fit between the

participant's values and characteristics and that of the research staff and research materials. Seventy percent of participants also discussed the importance of recognizing heterogeneity within cultural groups. They explained that taking an individualized approach, without compromising the science, may allow researchers to acknowledge individual experiences. Other solutions include collaboration with interpreters to provide services to non-English-speaking prospective participants and/or providing options for in-home or remote visits to overcome linguistic and physical access barriers, respectively.

Adopting a Flexible Approach to Recruitment and Data Collection

Seventy-five percent of the participants endorsed the importance of flexibility for the successful recruitment and retention of diverse groups. Participants frequently described recruitment strategies evolving as studies progressed. Recruitment techniques were incorporated or abandoned in response to study needs, and changes were guided by community representatives and relevant stakeholders. This adaptability extended beyond recruitment. For example, flexibility at the time of data collection was reported as necessary to retain participants (50 percent), particularly those with limited resources or constraints on their time due to competing demands (e.g., childcare).

Building a Robust Network by Identifying All Relevant Stakeholders

Eighty percent of the participants discussed the importance of identifying all stakeholders, highlighting the major role they played in informing study design and driving recruitment and retention of diverse participants. Who is considered a stakeholder, and their level of involvement, varied based on cultural preferences of the prospective participants, the condition being studied, and the nature of the research study. The term *stakeholder* was defined broadly to include caregivers, family members, friends, clinical providers and administrators, community advocates, peers, religious leaders, and political figures. Developing relationships with caregivers and family members was identified as instrumental to recruitment and retention of underrepresented groups. For example, "I realized that not talking to caregivers was a pretty big misstep in our original trial. If you have these populations that are vulnerable enough to have caregivers and other people who are already kind of with them maybe consider including them as part of the trial and obviously with patient consent, sort of incorporating it." Community advisory boards and other strategies for eliciting commiunity expertise were crucial to protocol development and study execution emerged in the context of this discussion. Finally, conceptualizing study participants as partners in research was an important component that required openness by the study team to learn from the participants' experiences.

Navigating Scientific, Professional Peer, and Societal Expectations

In the context of this theme, the participants described challenges related to acknowledging scientific and societal expectations while striving to maintain scientific rigor. Many participants perceived that efforts to promote representativeness, and decisions made to support these efforts, are not entirely appreciated by peers and organizations responsible for making funding decisions and/or budget decisions. Twenty-five percent of the participants described how creative strategies designed to engage communities that have traditionally been excluded from research are evaluated relative to more traditional strategies, which tend to be rigid. Thus, researchers are encouraged to use traditional approaches to retention and recruitment, which may be burdensome for prospective participants with multiple vulnerabilities, and may result in less participant diversity.

Another subtheme revolved around the incongruence between current emphasis on recruitment and retention of diverse participants and the consistent underfunding of researchers applying for grants to conduct this work. For example, "It seems that there's a real incongruence where the NIH is saying disparities work, disparities work, disparities work, and then you put it in and reviewers don't acknowledge the disparities aspect. They are fixated on errors in your approach or concerns about your theoretical model, and so it does seem that there is an incongruence in the way that the funding source of NIH wants to value efforts to recruit and retain these folks and then the way that it's reviewed. So that is an issue." Participants emphasized that efforts to be intentional and plan ahead to prepare for additional costs related to this work are undermined due to budget constraints. Seventy percent of the participants suggested that funding agencies, as well as those responsible for approving proposals and distributing budgets, should be required to gain competencies in nontraditional methodological research approaches.

Optimizing Study Team to Ensure Alignment with Research Goals

All of the participants described the composition of the study team as an important component of representative research. Study staff interact with potential study participants and are instrumental in recruitment and retention success. Diverse study teams were generally described as being helpful to recruitment given congruence between staff and potential participants, and this congruence was described in different ways depending upon the focus of the study (e.g., age, sex, race, ethnicity). Of note, 25 percent of participants added that cultural and linguistic congruence with the target population was not enough. That is, commitment to the study and its outcomes were as important when working with diverse communities. Retaining study staff over time was emphasized as very important to recruitment and retention success; however, this was also described as a challenge given issues with staff salaries.

Attaining Resources and Support to Achieve Representativeness

A variety of resources are needed to accomplish the goal of a representative sample. Eighty percent of the participants considered time and money as ultimately the most instrumental material resources necessary to conduct this kind of work successfully. With this in mind, funding support for these recruitment efforts was a main focus for specific funding announcements focused on underrepresented groups, expanded budgets for teams attempting to recruit and retain these groups, and flexibility within budgets to allow for deeper engagement of community partners. In addtition to funding, participants emphasized education of researchers and supports in the form of professional networks and institutional resources. Finally, material support for community organizations so that they can build infrastructure also emerged as part of this theme. In particular, resources that could assist these organizations in building the foundation for research would bolster these efforts for successful partnerships.

DISCUSSION

Employing our unique SRQA mixed-methods approach for this study, we examined facilitators to recruitment and retention in clinical trials. We conducted 20 in-depth qualitative interviews with researchers from U.S.-based studies who succeeded at recruiting diverse samples. In addition, we examined reporting practices of cultural and demographic sample characteristics in 162 randomly selected U.S.-based clinical trials published between 2001 and 2021. This approach provided the opportunity to learn about the unique and innovative techniques being incorporated by study teams nationwide, regardless of study size and/or national recognition.

This qualitative work sought to elucidate novel recruitment and retention strategies incorporated by researchers who have been successful at achieving representativeness in their cohorts. Several themes emerged that revolved around having intentionality and agency, building trust, recognizing heterogeneity, adopting a flexible approach to recruitment and data collection, and appreciating stakeholders. The characteristics of the study staff figured predominantly in the discussions, and involving and retaining experienced study staff was identified as a key ingredient to success by most of the participants. Barriers to recruiting and retaining experienced staff included low pay, job insecurity, and devaluing of their expertise based on lack of formal training. The findings suggest that equitable distribution of resources must extend to all levels of the research with underrepresented groups, community organizations, community advocates, caregivers, participants, research staff, and principal investigators. A genuine commitment to recognizing and respecting the contribution of each stakeholder is needed for success and sustainability.

Overall, most of the participants emphasized that expedient time frames and budget restraints inherent in existing funding mechanisms through the NIH and other agencies (e.g., the R01 mechanism) are not adequate to support research that seeks to include underrepresented groups. To develop the infrastructure necessary to support these efforts, flexible funding mechanisms that allow for inclusion of community partners will be essential. Additionally, specific funding announcements with a focus on supporting efforts to recruit and retain underrepresented groups will be needed going forward. In the context of institutional support, many participants called for academic health centers, which play a major role in employment of community members, maintaining the health of communities, and providing outreach to communities across the country, to build relationships with diverse communities that would, over time, allow for systems-level barriers to be addressed and trust to be established. These results suggest that financial support must precede additional calls to increase representativeness.

All of the participants expressed enthusiasm for the increased focus on representativeness by professional institutions, funding agencies, and scientists. They reported that shedding light on these issues is instrumental for science to remain relevant and consistent with its fiduciary duties to taxpayers who ultimately fund, at least in part, these efforts. The study participants believe that while most scientists value representativeness in research, many professional peers, who review grants and make decisions about funding priorities, are not receptive to research proposals demonstrating flexibility and adaptability. As such, resources in the form of training, support, and increasing representation of scientists from diverse cultural and academic backgrounds in review panels and positions of power is instrumental to this effort.

These results align with findings from prior research focused on participant-reported barriers and facilitators to enrollment in clinical trials (Ejiogu et al., 2011; Ford et al., 2013; George et al., 2014; Gilmore-Bykovskyi et al., 2019). The model proposed by Ford et al. (2013), which was informed by Hispanic/Latina(o) and Black/African American adults, organizes the sources of racial and ethnic disparities in recruitment in clinical trials across three major areas: (1) characteristics of study processes, (2) characteristics of health researchers, and (3) preferences and attitudes of community members and potential trial participants toward clinical trials. This study extends beyond these three areas by capturing systems-level issues related to the characteristics and values of academic institutions, the need for increased funding support for these efforts, and recognition of the importance of representativeness in clinical trials as a social justice issue. In a recent systematic review, George et al. (2014) identified shared and distinct facilitators and barriers to participation in research among persons from diverse backgrounds. Overall, cultural congruence between the study and the target community, benefits to participation, altruism, and convenience were listed as major facilitators to recruitment. In addition to cultural congruence (i.e., how good of a fit between the participants' linguistic, racial, and ethnic background and that of the research staff), many of the participants reported that successful recruiters valued research and appeared uniquely invested in understanding the experiences of those living with a given condition. According to George et al. (2014), participants from diverse ethnic and racial groups reported distinct barriers to participation. Many of the participants in our study cautioned that, even when techniques are implemented a priori to address these barriers, flexibility is instrumental to this work, as heterogeneity exists between and within different groups.

Across most of the randomly selected studies, incomplete reporting of sample characteristics was observed. While 90 percent of the trials (n = 162) reported information on the sex/gender of their sample, none described the sexual orientation of their study participants. Furthermore, less than 50 percent, 30 percent, and 20 percent of the studies reported whether their participants self-identified as African American, Asian and Asian Americans, and American Indian or Alaska Native, respectively (see Appendix C-4). Information about the representation of Native Hawaiian or Pacific Islander was reported by less than 10 percent of the trials. Most of the trials included in this study (90 percent) did not report information about the socioeconomic characteristics, disability status, and living arrangements (i.e., homeless or not) of their participants. Issues around intersectionality, or whether participants reported belonging to more than one group that has been historically marginalized, were not explicitly reported.

While 88 percent randomly selected trials achieved success in recruiting a representative sample, 53 percent of these were determined to meet criteria based solely on recruiting females. Several participants indicated that inclusion of women in their studies was due more to factors such as age (Graaf et al., 2018), disease (Duma et al., 2018), and the nature of the research trial (Vitale et al., 2017) than to intentional approaches to recruiting women. Regardless of the reason for adequate representation of women in these trials, this remains an important finding and focus for future work. Although women live longer than men, they report an increased number of years living with functional limitations (Zunzunegui et al., 2015), and older women score significantly lower on most indicators of subjective well-being and mental health than their male peers.

Less than 66 percent of the trials included a robust breakdown of the racial representation in the sample in ways that supported benchmarking (i.e., an "Other" category was used to describe anything other than Non-Hispanic white). These results suggest that despite consistent efforts to raise awareness about the importance of recruiting representative samples, including recommendations by the NIH and FDA to consider fundamental causes of disparities in research, improvement in this area is imperative. Calls to recruit representative samples must be accompanied by clear guidelines on how to comprehensively collect such characteristics. Moreover, consistent reporting of comprehensive sample characteristics may require explicit recommendations from journals as well as national institutions and organizations.

Several limitations merit mention. First, the definition of "success" focused on meeting U.S. Census targets based on regional proportions of diverse populations. However, chronic conditions are disproportionately distributed across different cultural groups. In an effort to account for variability in base rates, we selected trials across six disease types in the United States. Future researchers should examine these findings in the context of an expanded disease criteria that includes suicidality, substance use disorders, and other neurodegenerative conditions. Second, regional representation for a given trial was determined using geographical filters included only in PubMed. Additionally, the corresponding author's information, which is used to populate regional information in PubMed, does not necessarily reflect the location of recruitment. This represented a challenge to our explicit intention to benchmark to county, state, and national data. To the extent possible, our research team confirmed whether recruitment site was different from the corresponding author's site, and studies were benchmarked using U.S. Census data from where the study was conducted. While thematic saturation was met for this specific cohort of researchers, we acknowledge that it is possible additional themes could emerge if time allowed for inclusion of an expanded cohort of more diverse study team members.

CONCLUSIONS

This study highlights several strategies to promote representativeness in clinical trials and provide practical and innovative recommendations for relevant stakeholders in the field: peers, journals, and funding agencies. Ultimately, efforts to improve representativeness must involve provision of financial resources for research teams, material and social support for community advocates and organizations, and education about the relevance of these efforts to scientists, community members, and allied professionals. Priority funding should anchor research activities on representativeness, with community stakeholders at the forefront of every consideration.

Acknowledgments

The authors would like to acknowledge the participants who gave so generously of their time and expertise. We would also like to acknowledge the contributions of the National Academies of Sciences, Engineering, and Medicine committee. We are also grateful to Michael Davenport, Mary Hitchcock, Nicholas Picanso, Payton Sheridan, Nicole Razdolsky, Vanessa Cannaday, Alexis Micale, Rodolfo Pena Chavez, Ryan Burdick, Celia Deckelman, Andrea Gilmore-Bykovskyi, and Michael Pulia for their assistance with this project.

REFERENCES

1. Haidich, AB, Ioannidis, JP. Patterns of patient enrollment in randomized controlled trials. Journal of clinical epidemiology. 2001;54(9). doi: 10.1016/s0895-4356(01)00353-5. PubMed PMID: 11520646.
2. Clinical Trials AReNA Clinical Trial Delays: America's Patient Recruitment Dilemma 2012 [cited 2019 December 5]. Available from: https://wwwclinicaltrialsarenacom/analysis/featureclinical-trial-patient-recruitment/.

3. Census (U.S. Census Bureau). Hispanic Population to Reach 111 Million by 2060 2018 [cited 2021 July 30]. Available from: https://www.census.gov/library/visualizations/2018/comm/hispanic-projected-pop.html.
4. Carratala, S, Maxwell, C. Health Disparities by Race and Ethnicity2020. Available from: https://cdn.americanprogress.org/content/uploads/2020/05/06130714/HealthRace-factsheet.pdf?_ga=2.194956205.1806611089.1627327183-1860413079.1627327183.
5. SAMHS (Substance Abuse and Mental Health Services Administration). Top Health Issues for LGBT Populations Information & Resource Kit. Rockville, MD: Substance Abuse and Mental Health Services Administration; 2012.
6. FDA (U.S. Food and Drug Administration). Enhancing the diversity of clinical trial populations—eligibility criteria, enrollment practices, and trial designs guidance for industry: Center for Biologics Evaluation and Research Center for Drug Evaluation and Research; 2020 [cited 2021 July 30].
7. Duma, N, Aguilera, J Vera, Paludo, J, Haddox, CL, Velez, M Gonzalez, Wang, Y, Leventakos, K, Hubbard, JM, Mansfield, AS, Go, RS, Adjei, AA. Representation of minorities and women in oncology clinical trials: Review of the past 14 years. Journal of oncology practice. 2018;14(1). doi: 10.1200/JOP.2017.025288. PubMed PMID: 29099678.
8. Gong, IY, Tan, NS, Ali, SH, Lebovic, G, Mamdani, M, Goodman, SG, Ko, DT, Laupacis, A, Yan, AT. Temporal Trends of Women Enrollment in Major Cardiovascular Randomized Clinical Trials. The Canadian journal of cardiology. 2019;35(5). doi: 10.1016/j.cjca.2019.01.010. PubMed PMID: 31030866.
9. McCarthy-Keith, D, Nurudeen, S, Armstrong, A, Levens, E, Nieman, LK. Recruitment and retention of women for clinical leiomyoma trials. Contemporary clinical trials. 2010;31(1). doi: 10.1016/j.cct.2009.09.007. PubMed PMID: 19788933.
10. Vitale, C, Rosano, G, Fini, M. Are elderly and women under-represented in cardiovascular clinical trials? Implication for treatment. Wiener klinische Wochenschrift. 2016;128(Suppl 7). doi: 10.1007/s00508-016-1082-x. PubMed PMID: 27655475.
11. Frew, PM, Saint-Victor, DS, Isaacs, MB, Kim, S, Swamy, GK, Sheffield, JS, Edwards, KM, Villafana, T, Kamagate, O, Ault, K. Recruitment and retention of pregnant women into clinical research trials: an overview of challenges, facilitators, and best practices. Clinical infectious diseases : an official publication of the Infectious Diseases Society of America. 2014;59 Suppl 7(Suppl 7). doi: 10.1093/cid/ciu726. PubMed PMID: 25425718.
12. Rosende-Roca, M, Abdelnour, C, Esteban, E, Tartari, JP, Alarcon, E, Martínez-Atienza, J, González-Pérez, A, Sáez, ME, Lafuente, A, Buendía, M, Pancho, A, Aguilera, N, Ibarria, M, Diego, S, Jofresa, S, Hernández, I, López, R, Gurruchaga, MJ, Tárraga, L, Valero, S, Ruiz, A, Marquié, M, Boada, M. The role of sex and gender in the selection of Alzheimer patients for clinical trial pre-screening. Alzheimer's research & therapy. 2021;13(1). doi: 10.1186/s13195-021-00833-4. PubMed PMID: 33952308.
13. Khan, MS, Shahid, I, Siddiqi, TJ, Khan, SU, Warraich, HJ, Greene, SJ, Butler, J, Michos, ED. Ten-year trends in enrollment of women and minorities in pivotal trials supporting recent US Food and Drug Administration approval of novel cardiometabolic drugs. Journal of the American heart association. 2020;9(11):e015594. Epub 20200519. doi: 10.1161/JAHA.119.015594. PubMed PMID: 32427023; PMCID: PMC7428976.
14. Census (U.S. Census Bureau). The American Community Survey 2005-2019. 2019.
15. Haidich, AB, Ioannidis, JP. Patterns of patient enrollment in randomized controlled trials. Journal of clinical epidemiology. 2001;54(9):877-83. doi: 10.1016/s0895-4356(01)00353-5. PubMed PMID: 11520646.
16. Tay, TR, Pham, J, Hew, M. Addressing the impact of ethnicity on asthma care. Current Opinion in Allergy and Clinical Immunology. 2020;20(3):274-81. doi: 10.1097/ACI.0000000000000609. PubMed PMID: 31850922.

17. Jamerson, K, DeQuattro, V. The impact of ethnicity on response to antihypertensive therapy. The American Journal of Medicine. 1996;101(3A):22S-32S. doi: 10.1016/s0002-9343(96)00265-3. PubMed PMID: 8876472.
18. Odierna, DH, Bero, LA. Systematic reviews reveal unrepresentative evidence for the development of drug formularies for poor and nonwhite populations. Journal of Clinical Epidemiology. 2009;62(12):1268-78. Epub 20090417. doi: 10.1016/j.jclinepi.2009.01.009. PubMed PMID: 19375890.
19. Glasgow, RE, Huebschmann, AG, Brownson, RC. Expanding the CONSORT Figure: Increasing Transparency in Reporting on External Validity. The American Journal of Medicine. 2018;55(3):422-30. Epub 20180720. doi: 10.1016/j.amepre.2018.04.044. PubMed PMID: 30033029.
20. Ix, J. H., Allison, M. A., Denenberg, J. O., Cushman, M., Criqui, M. H. Novel cardiovascular risk factors do not completely explain the higher prevalence of peripheral arterial disease among African Americans. The San Diego Population Study. Journal of the American College of Cardiology. 2008;51(24):2347-54. doi: 10.1016/j.jacc.2008.03.022. PubMed PMID: 18549921.
21. NIH (National Institutes of Health). NIH Policy and Guidelines on The Inclusion of Women and Minorities as Subjects in Clinical Research 2001 [cited 2021 July 30]. Available from: https://grants.nih.gov/policy/inclusion/women-and-minorities/guidelines.htm.
22. FDA (U.S. Food and Drug Administration). Establishing a list of qualifying pathogens under the Food and Drug Administration Safety and Innovation Act. Final rule. Federal register. 2014;79(108). PubMed PMID: 24908687.
23. Nazha, B, Mishra, M, Pentz, R, Owonikoko, TK. Enrollment of Racial Minorities in Clinical Trials: Old Problem Assumes New Urgency in the Age of Immunotherapy. American Society of Clinical Oncology Educational Book. American Society of Clinical Oncology Annual Meeting. 2019;39. doi: 10.1200/EDBK_100021. PubMed PMID: 31099618.
24. Moher, D, Hopewell, S, Schulz, KF, Montori, V, Gøtzsche, PC, Devereaux, PJ, Elbourne, D, Egger, M, Altman, DG. CONSORT 2010 explanation and elaboration: updated guidelines for reporting parallel group randomised trials. International journal of surgery (London, England). 2012;10(1). doi: 10.1016/j.ijsu.2011.10.001. PubMed PMID: 22036893.
25. CDC (U.S. Centers for Disease Control and Prevention). Leading Causes of Death 2020 [updated 2021-03-01T08:25:51Z]. Available from: https://www.cdc.gov/nchs/fastats/leading-causes-of-death.htm.
26. Frey, WH. Six Maps That Reveal America's Expanding Racial Diversity Washington, DC: Brookings Institution; 2019 [cited 2021 July 30]. Available from: https://www.brookings.edu/research/americas-racial-diversity-in-six-maps/.
27. Lewison, DM. Perhaps the most widely used regional classification system is one developed by the U.S. Census Bureau. Prentice Hall; 1997.
28. Hill, CV, Pérez-Stable, EJ, Anderson, NA, Bernard, MA. The National Institute on Aging Health Disparities Research Framework. Ethnicity & disease. 2015;25(3). doi: 10.18865/ed.25.3.245. PubMed PMID: 26675362.
29. Taylor, HA. Implementation of NIH inclusion guidelines: survey of NIH study section members. Clinical Trials. 2008;5(2):140-6. doi: 10.1177/1740774508089457. PubMed PMID: 18375652; PMCID: PMC2861770.
30. Riley, WT, Riddle, M, Lauer, M. NIH policies on experimental studies with humans. Nature Human Behaviour. 2018;2(2):103-6. Epub 20171122. doi: 10.1038/s41562-017-0265-4. PubMed PMID: 30662956; PMCID: PMC6338435.
31. Crenshaw, KW. *On intersectionality: Essential writings*: The New Press; 2017.
32. Ford, ME, Siminoff, LA, Pickelsimer, E, Mainous, AG, Smith, DW, Diaz, VA, Soderstrom, LH, Jefferson, MH, Tilley, BC. Unequal burden of disease, unequal participation in clinical trials: solutions from African American and Latino community members. Health Soc Work. 2013;38(1):29-38. doi: 10.1093/hsw/hlt001. PubMed PMID: 23539894; PMCID: PMC3943359.

33. Boyatzis, RE. Transforming qualitative information: Thematic analysis and code development. SAGE; 1998.
34. George, S, Duran, N, Norris, K. A systematic review of barriers and facilitators to minority research participation among African Americans, Latinos, Asian Americans, and Pacific Islanders. American Journal of Public Health. 2014;104(2):e16-31. Epub 20131212. doi: 10.2105/AJPH.2013.301706. PubMed PMID: 24328648; PMCID: PMC3935672.
35. Gilmore-Bykovskyi, AL, Jin, Y, Gleason, C, Flowers-Benton, S, Block, LM, Dilworth-Anderson, P, Barnes, LL, Shah, M, Zuelsdorff, M. Recruitment and retention of underrepresented populations in Alzheimer's disease research: A systematic review. Alzheimer's and dementia (New York, N. Y.). 2019;5:751-70. Epub 20191119. doi: 10.1016/j.trci.2019.09.018. PubMed PMID: 31921966; PMCID: PMC6944728.
36. Ejiogu, N, Norbeck, JH, Mason, MA, Cromwell, BC, Zonderman, AB, Evans, MK. Recruitment and retention strategies for minority or poor clinical research participants: lessons from the Healthy Aging in Neighborhoods of Diversity across the Life Span study. Gerontologist. 2011;51 Suppl 1:S33-45. doi: 10.1093/geront/gnr027. PubMed PMID: 21565817; PMCID: PMC3092978.
37. Graaf, R van der, Zande, ISE van der, Ruijter, HM den, Oudijk, MA, Delden, JJM van, Rengerink, K Oude, Groenwold, RHH. Fair inclusion of pregnant women in clinical trials: an integrated scientific and ethical approach. Trials. 2018;19(1):78. Epub 20180129. doi: 10.1186/s13063-017-2402-9. PubMed PMID: 29378652; PMCID: PMC5789693.
38. Vitale, C, Fini, M, Spoletini, I, Lainscak, M, Seferovic, P, Rosano, GM. Under-representation of elderly and women in clinical trials. International Journal of Cardiology. 2017;232:216-21. Epub 20170110. doi: 10.1016/j.ijcard.2017.01.018. PubMed PMID: 28111054.
39. Zunzunegui, MV, Alvarado, BE, Guerra, R, Gómez, JF, Ylli, A, Guralnik, JM, Group, Imias Research. The mobility gap between older men and women: the embodiment of gender. Archives of Gerontology and Geriatrics. 2015;61(2):140-8. Epub 20150617. doi: 10.1016/j.archger.2015.06.005. PubMed PMID: 26113021.

TABLE C-2a Starting with Intention and Agency to Achieve Representativeness

Subthemes	Selected Quotes	Gender/Role
Importance of intentional approach	"And so if you want to be inclusive, you need to then think about how many from that population you want to enroll and begin to work towards that goal. That's number one. So I think that goes into the framework of intentionality, right? We need to be intentional. We want to do this and want to be intentional about doing it.... I believe very strongly that many times we want to do this as an afterthought. So we didn't go into the study saying that we want to enroll this many African American, Latino people, but [should] make that as part of the initial goal."	Male, Study Investigator
	"But the number one principle I do think is intentionality. You have to want to do it because expediency will kick in that you need to close the study in one year and you want to get those patients enrolled. But I do think if you start to plan from the beginning to have an inclusive group, that's important."	Male, Study Investigator
Motivation to pursue representative sample	"I mean, NIH, when you fill out the RPPR and you fill out your little diversity table, it's always hard to put in those zeros."	Female, Study Investigator
	"I think the mandate... was so hard. But when we had to get underrepresented groups three out of every ten, we did it and otherwise we don't . . . it just feels really good, working hard to get underrepresented groups."	Female, Study Investigator
	"[I] try to make the sample representative of where I am at the time, doing the work and where I am, there's a high percentage of LatinX. We need them in the work ... to understand what's going on with them."	Female, Study Investigator
	"So women are somewhat easier to reach in that way than men who are traditionally a little bit harder but not with this ethnic group. It was like we won't know anything about women's health if we don't have more women in the study, meaning equal numbers of men and women.... So it's like you got to get invested enough."	Female, Study Investigator
	"It was motivated by having a representative sample and understanding how so many individuals are not represented, I just think that's really important. And I just think as the scientist, well, we have to do that. It's part of who we are and it's why, as I said, it's hard but it's so important."	Female, Study Investigator

continued

TABLE C-2a Continued

Subthemes	Selected Quotes	Gender/Role
Budget planning with recruitment in mind	"So it's really budgeting for time and effort of people who are not typically thought of in grants. But those are like most of my grants, like all of the funding goes externally like subcontracts to different partners."	Female, Study Investigator
	"When he was writing, he's like 'Send me the budget for the recruitment, for the outreach.' Amazing, I mean I put in there, I can tell you my budget was transportation to events and food for events, and he was like 'Great, thank you.' No questions asked, submit it."	Female, Study Coordinator
Proactive study design	"We said we were going to design the study to do subgroup analyses to look at if the intervention was effective and in historically underrepresented groups. And that was important for us."	Female, Study Investigator
	"And so we designed with and for people at risk for the worse outcomes . . . And so the way you design for that, a lot of people say, well, it's only among Black women that we're testing this intervention and that's great. I don't have a problem with that at all. But the way I have done it is to say we're going to ensure that there are enough Black women in the sample that we can do a subgroup test to make sure that the effect size that they see is on par with the effect size that non-Black women see to try to understand if our interventions are exacerbating disparities."	Female, Study Investigator
	"The other thing that I think is really important is designing and piloting study materials, whether it's the study intervention or the consent form or the recruitment process, for underrepresented groups, and then they will play fine for the other groups."	Female, Study Investigator
	"So the whole grant was written around community engagement, mixed methods, both quantitative and qualitative, really understanding segmented assimilation and new ways of thinking about immigrant health and how to really quantify where the gaps are, what the barriers are, and how we can improve health."	Female, Study Investigator
	"I think I would look pretty closely at the inclusion/exclusion criteria. I guess of the top of my head common inclusion/exclusion that would apply or would be different based on your gender or race would be a lot of rules around pregnancy and having to use certain methods of birth control. . . . I think that can deter some patients. And, on the flip side, if you're a male enrolling into a clinical trial, you don't have to have any sort of birth control."	Female, Study Coordinator

TABLE C-2a Continued

Subthemes	Selected Quotes	Gender/Role
Combining recruitment approaches to optimize enrollment of diverse groups	"We were purposeful in our recruitment strategies, we used EHR data to prioritize folks that were either from a minority racial or ethnic group or had an indicator for insurance that they may be uninsured or underinsured. And we prioritized recruitment of those groups."	Female, Study Investigator
	"We tried other methods, like using electronic health records using self-identified ethnicity that is collected by health systems data to reach out to this particular demographic that we really wanted to recruit. And those letters were just terrible in terms of getting any yield . . . it's horrendous because it's just not meaningful."	Female, Study Investigator
	"So basically how recruitment happened is based on individuals who were Medicare eligible, who lives in these neighborhoods . . . and so they were selected based on this random recruitment effort to find people in each catchment area based on the census in the neighborhood. And that was the goal to get this representativeness. So the beginning of this study, there is actually a lot more White people . . . But then that started to switch as the composition of the neighborhood switched."	Female, Study Coordinator
	"So I think in some sense the clinics did that for us, like if this is a clinic that largely serves the homeless population downtown and we partner with that clinic, we don't need to do a lot of extra stuff to reach those patients. So making sure those clinics were priorities for us and we did adjust a lot of our approach in working with the clinic."	Female, Study Investigator
	"So we partnered with community organizations, faith based groups, leaders in those vulnerable communities that have traditionally been left out of research and just did a lot of outreach activities, both in person and in terms of health fairs and other community venues."	Female, Study Investigator
	"So we're using probability sampling in terms of knowing the demographics from the last five years of the American community survey in these geographic locations, knowing the sex and age distribution and education distribution in these communities…So we're not doing all community engaged methods. We have to put some kind of boundaries around that so the validity of our data is still there and will not be questioned."	Female, Study Investigator

continued

TABLE C-2a Continued

Subthemes	Selected Quotes	Gender/Role
Designing marketing strategies to reach underrepresented groups	"So we used a lot of different community activities as well as a radio and ethnic radio station ads and interview with radio stations that are to reach these populations, some TV spots as well. A lot of faith based organizations helped us with providing us a little bit of advertising in a way but letting people know about the study."	Female, Study Investigator
	"The venues that we were allowed to use marketing in front of ethnic markets and restaurants and in other community settings. And a lot of that was actually quite successful in getting the word out and getting people interested."	Female, Study Investigator
Reciprocity with Study participants and community	"I also think it's important to share how this study is going. We put together a tipsheet because when you're doing this study it's like a black hole. So I'm in this trial and I have no idea what's happening. So I think getting an idea of how this study is going, regular information shared with people in the trial about the progress of the study is important."	Male, Study Investigator
	"We actually suggested providing other ancillary services, educational materials. So you're in a study, you know that these other health issues are related to X, Y, Z, so I think a lot of people select into those preferences and provides useful information for them to engage them in science without compromising their study goals."	Male, Study Investigator
	"One of the key challenges that we face is downstream care. Sometimes our goals are undermined by word of mouth, so someone gets a positive test and they can't do anything else with it and they say 'Oh yeah, I couldn't get that colonoscopy, they don't care about me' so I think that sort of downstream continuum of care, a cascade of care that is needed for the clinical trial needs to be provided. And to me that is best practice."	Male, Study Investigator
	"I think we need to develop a process by which we have relationships with people. It is through that ongoing feedback to the community or participants over time."	Male, Study Investigator
	"Community engagement is important for the downstream care after clinical trials—it is critically important and should not be ignored because I do think that, when we fail to do that, it is in the care process. A clinical trial undermines efforts to build trust and growth in science."	Male, Study Investigator

TABLE C-2a Continued

Subthemes	Selected Quotes	Gender/Role
	"And so we do try to give back. We don't just recruit, we always try to give back to the community. I think that's really important if you want to have a relationship with the community, you don't just take. Whatever that community is, we try to teach you, we go to health fairs, we try to give something back."	Female, Study Investigator
	"Tokens of appreciation, we were constantly giving feedback to the clinic about how many people we were recruiting, and then we gave feedback on our results and things we were finding and publishing."	Female, Study Investigator
	"That's who we're recruiting are staff members and people interacting with participants from the community. So that's another really big way that we're invested. It's that reciprocal relationship."	Female, Study Coordinator

TABLE C-2b Establishing a Foundation of Trust with Participants and Community

Subthemes	Selected Quotes	Gender/Role
Importance of building trust with community and prior issues of distrust	"And the reason that I don't think it worked well [sending letters] is that there's no trust. There's a lot of mistrust in getting a letter from a random person, even though it has a university letterhead on it. You don't know anything about the person or the research or if you're undocumented or don't speak the language, if you've never been exposed to research, what the point is for research, there's many layers of trust that cannot be broached with an invitational letter and brochure."	Female, Study Investigator
	"I trust [her] but do I trust the system? Do I trust the hospital? ... I have some case studies and that actually comes from you showing them what you are doing with the data and what is being not only done but not done. Are you giving feedback individually? Are you giving feedback as a whole? It's the community, how is it being used to further policies?"	Female, Study Coordinator
	"So I would give a talk and try to sit with people. And we had food afterwards usually, so we could all just sit and talk casually. But they're telling me, over and over again, there's just a lot of distrust in the medical community and I get it, I understand why."	Female, Study Investigator
	"This one community that I'm thinking about has been a little historically suspicious because of bad experiences they've endured of medical research and perhaps academic medical research and so sending out a single notice is not going to be sufficient in order to have meaningful recruitment of these groups. It's really going to start with building relationships of trust and then later availing those groups of opportunities."	Male, Study Investigator
	"So I think it is the relationship and trust to me is the key. Once trust is established, people will do things that I believe are coming from you and you better keep that promise."	Male, Study Investigator
Dedication to true engagement with community regardless of study enrollment	"And so I think that's a way to cause [distrust], you have these studies where people are meeting that requirement and they're not treating the community very respectfully often or they just don't know how. And it's maybe, it's usually unintentional but it's a consequence."	Female, Study Coordinator
	"I don't know if this is tested anywhere is this idea of helicopter history. So we're coming to do our research, we're done, and then you're never seen again. Then the next time a research study is done, you can see it and it's done and you never see it again. I think we need to develop a process by which we have relationships with people."	Male, Study Investigator

TABLE C-2b Continued

Subthemes	Selected Quotes	Gender/Role
	"It's really hard work in terms of it takes a lot of energy and time investment. You have to really stay connected with the community or you cannot just go in and out. I mean, that's a commitment, right?"	Female, Study Investigator
	"We participate in community events that the clinics did. So if they did a diabetes day or a health fair, we were there with our table and we didn't really recruit people from that but it was just sort of part of being in the community we helped with. . . . Resources that the clinics have put into place, we were able to participate in, and that helped us as well."	Female, Study Investigator
	"So the key is as a study team we need to also be doing community outreach and service to actual caregivers and [patients] to make that connection."	Female, Study Investigator
	"Having staff available to go to those satellite sites as needed I think is a good strategy to maybe improve recruitment of specific populations."	Female, Study Coordinator
	"We were trying to be visible and physically present as often as possible within the clinic and also really so that the staff were very familiar with us, they saw us. So we wanted to be present and we didn't want to disappear once we start the study or the data collection. I think that's really important, wherever you could be, whenever you could be physical, and let them know that you are still here."	Female, Study Investigator
	"I think and it may not be that the people in the clinic are not necessarily are participants but I think I'm talking about it's more a long-term strategy that is not just for your study. But I do want to maintain this relationship and I want to continue to recruit diverse patients and families in our study that I have to be present. I think the study team has to be present in many different ways."	Female, Study Investigator
Bring research to the community	"So really embedding our staff in the communities, completely doing all of the outreach and all of the clinical exams in the community setting, making it as less clinical as possible, making every attempt to reach people where they live, work, pray, and play."	Female, Study Investigator
	"Right now, the series that we're going to start, we do them in the area that we recruit from. So I also like to detach that from the institution because we have institutional events maybe here because it's easy. But I'd rather find places for me to host events that are outside of the institution because those are more safe. I think those are safe spaces."	Female, Study Coordinator
	"We have to provide the best care and the best trials where people live and minimize the disruption that they face. And only until we do would we see sustained improvement in access to different clinical trials."	Male, Study Coordinator

continued

TABLE C-2b Continued

Subthemes	Selected Quotes	Gender/Role
Developing lasting relationships with study participants	"[We are] trying to be much more centered on that person and their individual needs."	Female, Study Investigator
	"And the women are probably just overwhelmed with other work that they have to do in the home with childcare, with employment, with finances. So making it as easy as possible for women. We have had our coordinators actually go to the woman's house, picked them up and come with them either on public transportation or a shared ride or whatever, just kind of building more of a relationship. So they feel like they know this person. They feel safe with this person."	Female, Study Investigator
	"We send them birthday cards and holiday cards. We do all the obvious things that many, many cohorts do. But we try to always put more of a personal touch. So the coordinator that they know signs the cards with their name so that they know the signature of the coordinators. They know that she really did that. On Monday, when they call they want to talk to the coordinator. They know her. They trust her."	Female, Study Investigator
	"Yeah, incentives, we paid them. And then establishing that personal connection with them because they were letting us into their homes with these video recorders and things. So I would talk to them on the phone each week. And sometimes these conversations would last 15 minutes, sometimes they would last 2 hours. Where we would just chat about 'How's it going?' I really tried to get to know them on a personal level."	Female, Study Investigator
	"Just sitting in the church cafeteria or wherever we eat and just sitting and breaking bread with people. That's just a traditional time for people to maybe gain a little bit of trust. And somebody would ask me a question that they didn't want to ask in front of everybody. So I'd go at this table with three people and we'd talk and it was just so much more intimate. And I think they felt listened to and asking questions that they didn't want to ask in front of everybody, I think. And by the end, I knew their names. They knew my name and I did get a lot of calls."	Female, Study Investigator
	"The project coordinator and the recruitment coordinator were two different people. And it actually helps . . . because then I would become the person who can be neutral and because I'm not calling to recruit you. I'm actually telling you about the study. And I'm going to listen to you. And you're going to decide based on what I'm saying whether you want to participate or not . . . because you're now somebody who is more likely to be trusted because you're just there to explain and you can maintain a relationship."	Female, Study Coordinator

TABLE C-2b Continued

Subthemes	Selected Quotes	Gender/Role
	"What I did find is that sometimes it's too much. You have to make yourself available but also like, just lean back, because I do find that unfortunately this population of elders, there's a lot of people trying to get them. So I have found [with] . . . this community, less is more. We don't send cards, I know a lot of people do but the birthday carts that to them, believe it or not, . . . feels a little like, oh, it's just random. Like everybody, like another thing…doesn't cut it. What we do is . . . we find anything that's awesome about them. They could be a painter. They could play guitar. They could be clay sculptors. Whatever it is, we find out. I have a section where I write that little tidbit about them and we'll make sure when we see them again or talk to them or I call them and I'll say something. Something's interesting, there's something personal about it. Personalized, no cookie cutter response that's done to everyone."	Female, Study Coordinator
	"That's why we don't force, everything's voluntary. . . . They can withdraw at any time. So we make sure that they instill that in anything that we do, no forcing answering questions. Their well-being is first, the study goes second. And then it just always comes first with us because we just put them first. So they put the study first."	Female, Study Coordinator
	"I mean we do provide incentives for follow-up interviews. I think that certainly helps. But I think once you have established a relationship through your first baseline visit."	Female, Study Investigator
Maintaining a favorable study reputation with the community	"And so that's why we're hoping that the use of these community outreach events that we did and the people who care about us and see a face to a name and you hear about it from other trusted leaders and champions in the community and that would give us an air of legitimacy in reaching out."	Female, Study Investigator
	"It starts conversations with friends, conversations when we're not present, which is in the kitchens and the dining rooms. So then we find a way for our findings and what we do to become part of their everyday conversations. So when we started doing that, then we started getting people who not only were aware of studies but we started getting people who knew of other people who were in research studies but had no idea that's what it was."	Female, Study Coordinator
	"But how do I make sure that the same institution that they work for also doesn't screw me over for your study. So that's just to tell you how the instances, even within the research community, impacts other studies indirectly, maybe not for the long run but it does for now."	Female, Study Coordinator

continued

TABLE C-2b Continued

Subthemes	Selected Quotes	Gender/Role
Developing lasting equitable relationships with community partners	"If we activated communities enough to buy into the concept of research and they are advocating for research, then all of this will kind of change, right?"	Male, Study Investigator
	"We try to go to the community like beauty shops, barber shops. Those are really good we found to post things because it's pretty small. And we can sometimes, if they want to, they can post something right up on the mirror at the beauty shop. So we do a lot of churches and go to church events. And recruit there."	Female, Study Investigator
	"When they need something, like they're having a health fair and has nothing to do with the research we're doing, they need to reach out to me. And my academic institution has plenty of people who can volunteer at their health fair and can be there and can partner and do things if they need help with. So it's not only about us and the research we bring to them but it's about they want to get this health care done. Got that kind of relationship."	Female, Study Investigator
	"I'm also going to organizations that have budgets or . . . do these events within the community. And I'll just circle in . . . finding those organization that are not necessarily mirroring what you're doing but finding connections because everything has a connection, right?. . . Although there will be limited funding and strategies but I think there's always a connection with what you do within any community. There's a way for you to connect and then bring your message."	Female, Study Coordinator
	"The other resources were there was a consortium . . . here that all of the . . . clinics would meet once a month. And so that gave us opportunities to be present, understand initiatives that are going on, present the study, present findings from the study. . . . If it wasn't for that infrastructure, it would have been much harder to sort of roll out a new . . . clinic to build new relationships."	Female, Study Investigator
	"It is a very slow process. So we started with one clinic that we had done a lot of formative work and we partnered with them for some time."	Female, Study Investigator
	"It's a two-way street. I don't just go to them when I have a study. And I can't expect them to be open and ready to help me with every study and I'm not truly there for them. So it's not only me, but it's like having this kind of relationship that is enduring and takes time to build. And it's not a trivial commitment. It's a real long-term commitment. And so we built these relationships with our community partners for now more than a decade and have been and those relationships come with both give and take of information."	Female, Study Investigator

TABLE C-2b Continued

Subthemes	Selected Quotes	Gender/Role
	"Our academic partners have been working with these community organizations and actually have community health workers who worked with them on other projects. So it's easy to take them from one project to another until they have this track record. And it works really nicely for them because they have built in trust already."	Female, Study Investigator
	"And I have maintained relationships with people for years without them actually joining the study. And one of those people, she is a community activist advocate. And people thought that this person has been studied for years. And I'm like, no, she just joined two years ago. They're like, 'What?' I'm like 'Yeah because it just took that long.' It may have to for whatever reason but it took that long for her to say yes."	Female, Study Coordinator
	"We have mature community engagement programs . . . I choose to hold a community partnership. Recruiting within that partnership is easy because you have trust build over many years."	Male, Study Investigator
	"I think that is the goal to get to full equity with the community partner, writing grants and getting the money and sharing everything from the ground up to the study. I think we're still unequal with academic partners doing grant writing, getting funding, and working with community partners and giving them funding from the grant. So I think there's still this hierarchy, unfortunately, we're trying to break those down. We're trying to get to as much parity as possible and that's just going to take time and it's going to take investment."	Female, Study Investigator

TABLE C-2c Anticipating and Removing Barriers to Study Participation

Subthemes	Selected Quotes	Gender/Role
Increasing physical access	"In my view, it involves some remote access to trials . . . so that not everything needs to be face-to-face. If you put transportation between you and a trial, it falls down [in the] engagement of patients"	Male, Study Investigator
	"Paying for shared rides, Uber or Lyft to make it easy and convenient for people to come instead of having to pay for public transportation."	Female, Study Investigator
	"So that's one of the ways, a lot of people as they get older and more frail, they don't want to travel into the clinic appointments and do all these tests. So we go into their home doing what we can in the home and getting some measures rather than all of the measures."	Female, Study Investigator
	"Their coming into the clinic like three days a week to get . . . lab samples and that is a lot of driving, that's a lot of time to . . . have to take off work, or have to take away from family. And not all patients are privileged enough to be able to take time off and come to the center every day."	Female, Study Coordinator
	"And so travel to centers . . . it's a big barrier . . . so assisting in transportation centers is important if that's required. Remote monitoring is important because I think why bring people back just to check that they're ok when it can be done remotely."	Male, Study Investigator
Increasing linguistic access	"If you want to get folks involved in your research and they happen to be part of that community, boy, you better have people on your team that have language skills related to that and certainly you better have your documents professionally translated into those languages."	Male, Study Investigator
	"So there were two language translations that were required in order to do our study . . . if you don't have those materials prepared and you don't anticipate the need to have those materials a priori, it sort of becomes a self-fulfilling prophecy in that you're not going to accrue well or at all in those populations."	Male, Study Investigator
Ensuring study team accessibility	"We provided a helpline where people could call and just leave a voicemail. And there was no threat that anybody was going to answer that they would have to speak to. So they would just call the voicemail and say 'Hey, my cell phone number changes. Here it is.' or 'Hey, I'm moving. Here's my new address.' And we got a lot of sort of contact change information from that helpline."	Female, Study Investigator
Recognizing within-group heterogeneity and tailoring approaches	"I think it would be much more important if you did include the actual ethnic groups that you're asking about. You can't generalize what I said about all the studies we've done to other groups. . . . None of these lessons learned naturally translate to other groups."	Female, Study Investigator

TABLE C-2c Continued

Subthemes	Selected Quotes	Gender/Role
	"So it's about adjusting and making changes based on the observations of people who are having one on one contact with the cohort within the community."	Female, Study Coordinator
	"So again, I can tell you how it works here. It doesn't necessarily mean that it's going to transfer to another community… Maybe that is a strategy, to look for diversity even within the groups and see what works for each group and take the time to do that and do exploratory findings."	Female, Study Coordinator
	"There was no cultural tailoring at all. There was a ton of individualized tailoring. The intervention itself is highly individually tailored. And so we just developed personalized approaches to everyone. And, in doing that, we didn't have to put people into categories to try to tailor to them."	Female, Study Investigator
Adapting study materials and consent process	"You have to spell it out. . . . And how do you explain this to persons who don't have a background on just simple science? Let's say because these people have low literacy or didn't go to school for many years. Well it just takes time. . . . I tend to have conversations with them . . . it may make sense to me but I may not be explaining it well and what does this mean to you?' And then with that feedback, give some suggestions back to the PIs and then we make those changes to the consent form."	Female, Study Coordinator
	"Consent process was long. It was actually very well written but I can imagine people would say, well, I'm not understanding this concept, even if you translated it into another language people could not read. And then they will not allow you to read it to people. So you have a process that becomes very difficult. So I think the consent process is harder than it needs to be. And being accessible, in my view, is some of the ways that we've sought to [overcome] that."	Male, Study Investigator
	"And so in our part of the country, persons don't always have the best education . . . some have like third grade education. And so we just try to use the simplest words . . . we try to use . . . very few direct words. And we spend hours, we revise and revise to get it to that point."	Female, Study Investigator
	"Our materials do have diverse people on them, all kinds of LatinX and Asian and African American because I think that's important. I mean, it's so important. What you see is that 'I don't see me there'."	Female, Study Investigator
	"If a patient is deaf or blind, just having those resources available in our center when needed so we are not limiting our recruitment of 'disrepresented' or misrepresented patients in any way."	Female, Study Coordinator

TABLE C-2d Adopting a Flexible Approach to Recruitment and Data Collection

Subthemes	Selected Quotes	Gender/Role
Adapting recruitment approach to address low enrollment	"We applied an agile process, we constantly looked at data, we were constantly saying "Is it possible to get the sample we want from this clinic? Is it our processes that are the problem? Is it the patient pool that's the problem?' And through doing that we engaged with the clinicians there . . . saying help us crack this nut. And they were like 'You know, you're not going to get it there. You need to be looking here'."	Female, Study Investigator
	"Now in subsequent waves of recruitment, we've used a lot more community engagement."	Female, Study Investigator
Adapting study protocols as needed	"I guess one of the models that I do a little bit differently than some of my colleagues. . . . I meet with the entire team once a week and I also meet one on one with people. . . . I got to hear them saying 'This part of the protocol is making people sort of turn off.' And so I was like, ok well, then let's revise that part of the protocol."	Female, Study Investigator
	"So one clinic didn't have space for us to meet one on one with people. And so we would try to figure out, is there a safe way? Like, for instance, is it safe for a research assistant to meet at the mission at the homeless shelter next to the clinic? Is that a good idea or not? So we did a lot of tailoring to our processes based on the clinic requirements and restrictions."	Female, Study Investigator
	"We allowed women to bring their spouse to the visit and then allowed the spouse to have a limited exam. And we didn't use the data for the spouse. So we got her approval to have a limited exam visit for the spouse to be included because the spouse would bring the women to the exam and the woman is a participant. It helped us retain women in the study."	Female, Study Investigator
Flexibility in data collection procedures	"We have to meet with them weekends and whenever we can. Yeah, that's one of the criteria for being able to work on my project, just be available . . . So almost everyone has some degree of expectation of flex schedule."	Female, Study Investigator
	"So we had to be very flexible in how we collect the data. We ultimately ended up giving people multiple data collection options, so we tried to enroll everyone and do baseline data collection in person for folks, for literacy reasons, for understanding comprehension and for trust building. And then after that, they could meet us in person or in the clinic. They could meet us in person in our research offices. They could do it online via REDCap. They could do it via phone with a research assistant. They could be mailed a paper survey. And similarly they could go in for a . . . test at a clinic or they could do a mailing kit."	Female, Study Investigator

TABLE C-2d Continued

Subthemes	Selected Quotes	Gender/Role
	"I get pushback from people on other studies that I'm not a PI when I describe that we use all these different methods for data collection. People react to that and I disagree. I think data are far better than no data because, when we have no data, it's biased in systematic ways. When you have data from multiple sources, it's just a little bit more variable. And all that does is make it more difficult for you to detect, well potentially we don't know for sure, an effect but it doesn't systematically bias the effect that you're going to find. And so I just think that it's important to offer as many forms of data collection as you can to increase retention, even if you have some measurement error."	Female, Study Investigator

TABLE C-2e Building a Robust Network by Identifying All Relevant Stakeholders

Subthemes	Selected Quotes	Gender/Role
Caregiver and family involvement in study helps participation	"Having a family member . . . or someone who helps them with their day-to day tasks, that was extremely important for patients, and, perhaps, if they didn't have that in their life, it would have been difficult for them to enroll and complete this study."	Female, Study Coordinator
	"A lot of times patients rely on family member, or friends, or other people in their lives to get them to appointments for this study."	Female, Study Coordinator
Participants as partners in research	"And also revising things not just on our feedback but their feedback. So, for example, we had administered a discrimination index questionnaire. So one of the causes or reasons for discrimination did not include gender. I think it was gender orientation. It was one participant who was like 'You're not including this.' So, we gave feedback to the PI who had, she's the one who had put the question in. And Dr. X had a very long conversation with this participant about how this was not capturing a reason why it was discrimination and, guess what, we went and revised that just because one participant said it. So it's so important that they're listened to, that questions are being asked, but they could be revised because it's listening to their experience."	Female, Study Coordinator
	"I also think X does a really good job of, when they're in the study, making them feel like they're part of the study . . . like with the feedback and really taking it in. And they really feel like, ok, you're not just here like with your ivory hat on and telling me what to do. I actually feel like I'm a part of this study and I've been here for twenty six years doing this, like I feel like this is my family."	Female, Study Coordinator
	"So older adults, the participants, were definitely the key primary stakeholders."	Female, Study Investigator
	"But I think that actually giving people the opportunity to give their feedback, turning them into sort of active participants in their own intervention of things, I think there's a lot of power to that."	Male, Study Investigator

TABLE C-2e Continued

Subthemes	Selected Quotes	Gender/Role
Staff as partners in recruitment	"And so one of the things that really helps is that research assistants who are on the ground going into clinics got to hear me think through scientific decisions and say, 'I don't think that's going to work in this clinic. Here's why.' Or I got to hear them saying, 'this part of the protocol is making people sort of turn off.' And so then let's revise that part of the protocol. So it was way more of a free flow of information from the boots on the ground people to the decision maker for study design. And through that process they understand a lot of why we were doing things a certain way. And I understood when that way wasn't working and could make changes to it."	Female, Study Investigator
	"You have a team where people's, my, input is heard, which is . . . not common, right? Just because I don't have a PhD background but I am well-versed in what I do in recruitment."	Female, Study Coordinator
	"So again, it's about adjusting and making changes based on the observations of people who are having one on one contact with the cohort within the community."	Female, Study Coordinator
Community members inform research procedures	"Get the input of those who are actually working within the communities. . . . I think you will come up with a lot of different ways how . . . to diversity their cohort."	Female, Study Coordinator
	"So we have community advisory boards that are built very early in the process and each site has a different community advisory board because the issues that come up with each geographic location are very different and the communities to serve are very different. . . . We try to get a good representation of age and gender and different types of work and the experience in the community."	Female, Study Investigator
	"You would go to the community . . . and say 'I have an idea for research. I'd like your opinion on what the community might feel about this. Am I trying to get too many people? What would I need to establish a relationship? How can I help you to help me hire out of the community so that they can have people that are easily accessible to ask questions?'"	Female, Study Coordinator

TABLE C-2f Navigating Scientific, Professional Peer, and Social Expectations

Subthemes	Selected Quotes	Gender/Role
Inadequate understanding of recruitment and retention challenges among proposal reviewers and funders	"A barrier to that is the . . . misinformed notion that you have to be powered for an interaction term and that's a real problem in review because it's not actually what you want to know. You don't want to know if the intervention was significantly differently effective. You want to know, was it effective in minority groups? Was it effective in low SES groups? And so I think getting that communication to reviewers is going to be important for this work moving forward."	Female, Study Investigator
	"I think heavier weight in review. Instructions for reviewers that heavier weight should be given to plans for these types of recruitment efforts, that things like measurement bias should be downweighted and it seems that there's a real incongruence where the NIH is saying disparities work, disparities work, and then you put it in and reviewers don't acknowledge the disparities aspect. They are fixated on errors in your approach or concerns about your theoretical model and it does seem that there is an incongruence in the way that the funding source of NIH wants to value efforts to recruit and retain these folks and the way it's reviewed. So that is an issue."	Female, Study Investigator
	"The funding agency, if anything, it's been a barrier because of the reviewers we've gotten. They don't understand the significance and we have to basically turn some results to have them. But we've had a lot of biased reviews. Each report that we've written that has been funded has taken multiple, multiple tries. . . . So it's not very easy. We're pushing, it's an uphill battle every single time."	Female, Study Investigator
	"They have to be mindful of it. They have to have representative review panels. Panels cannot be a single demographic, mostly white male groups. You have to have representation of the communities that are represented in the US on your panels. You have to educate people about the need for this."	Female, Study Investigator
	"And that means that the panels that review the research have to be educated and representative and cannot be biased the way they currently are."	Female, Study Investigator
	"I do think that more people from the NIH need to come to low income areas, urban areas, and do home visits for like, spend a day visiting families . . . because when you start visiting families and start seeing what's really the issue, it's hard to ignore it."	Female, Study Investigator

TABLE C-2f Continued

Subthemes	Selected Quotes	Gender/Role
Shift from historical perspective on clinical trial recruitment to focus on diversity	"I think if I compare my thought process X years ago when we started the study and now, my single thing I would point to that's different is my awareness of underrepresented groups has been much heightened. And I don't think that I have to go through all the reasons why that is because I think you know what those reasons are anybody who's sentient and keeping track of the current events of the day would have some idea why that would be. And so, because my awareness has been increased, I don't think that's peculiar to me being a scientist as much as it is just being a member of the human race in the cities in the US in the current times."	Male, Study Investigator
	"I think it's more of like the scientific rigor about recruitment, but I think this cohort in itself is unique because it was diverse before it had to be diverse. So now you have the whole NIH diversity initiatives and I can tell you that if someone is doing it for those reasons, it is going to show because it's already showing in many other studies."	Female, Study Coordinator
	"They are specifically asking about how many patients of a specific ethnicity or race you think you could recruit at your site. So a few years ago they would never get as specific with that. They would just ask total patients that you think you could recruit. But now they are specifically focusing on improving clinical trial diversity."	Female, Study Coordinator

TABLE C-2g Optimizing Study Team to Ensure Alignment with Research Goals

Subthemes	Selected Quotes	Gender/Role
PI investment in supporting and training study team and leading by example	"I guess one of the models that I do a little bit differently than some of my colleagues is that I imagine a lot of people that run trials as PIs meet with their coordinator and their coordinator meets with the team. I don't work like that. I meet with the entire team once a week and I also meet one on one with people. But the coordinator doesn't like to filter the information up to me or down through her."	Female, Study Investigator
	"I do vividly remember training them in how did you inform consent in a conversational way and like looking the person in the eye and not having to read every word. And how to do teach back in a very casual way. . . . And so that process was important as well, especially when they're not in one research office, they're out in the community."	Female, Study Investigator
	"We keep track of people are doing that, we have accountability checks. So if somebody hasn't been following up with their folds for retention, they sort of appear on this slide in front of a team and it's like 'Do better' and we'll move on with our lives."	Female, Study Investigator
	"I do like staff building events. Like I have people over at my house. We do a lot of . . . family building events like gifts, but not because I want to just retain them, because I really do care for them as human beings. They're part of my work family. And so we're on texts together and we know each other now. And it's been a long-term relationship that we built in years, trying to drop out all of the formality from day one and to make this like a group effort. And so we're in this together."	Female, Study Investigator
	"I felt the obligation that I needed to lead by example. So, to be honest with you, that maybe that was a prideful reason for why I did what I did. But that was the motivating factor. Day after day, week after week, year after year, the study was, I've got to set the example . . . I mean, if you're going to be a leader, you need to be a leader in all aspects of things."	Male, Study Investigator
	"So sometimes trying to create that kind of environment on the team where that's a topic of discussion. It's prioritized as something that's important to us. It's kind of like that atmosphere type of thing."	Female, Study Investigator
Expressing trust and appreciation of staff	"I also do a lot of work with my team to teach them that my priority is them and then the participant and then the data."	Female, Study Investigator

TABLE C-2g Continued

Subthemes	Selected Quotes	Gender/Role
	"So having the same staff at our site, we've had the same staff for 11 years now and are so thankful and grateful. And we've done everything to retain the staff . . . because they're the face of the study."	Female, Study Investigator
	"Well you had mentioned salary in the context of health care coordinators. I'm not sure my challenges with the turnover of the study's personnel were related to salary as much as it was related to embracing them, making them feel like they were part of the team, help recognizing their important contributions."	Male, Study Investigator
	"Like all the people of color on my team are making the least amount of money and all the white people were making the most amount of money. And so then it's all those things I realized over the years that we just don't place the same value on, like being an amazing recruiter and an amazing person for retention."	Female, Study Investigator
	"But I feel like the staff are the experts when it comes to the patients. I mean, I really do rely on the staff. And there are also, again all of their staff are awesome and great at explaining things."	Female, Study Investigator
Consistent and experienced staff	"We try to have the same person reach out to them to collect data. . . . It helps to have a consistent person, like, 'Gee, you know, XXX just called me again. I know XXX. I recognize XXX. That is nice. I like XXX. And I'm more likely to pay attention.' So I think a consistent individual to follow through on multiple contacts without . . . making yourself just a nuisance. And I think those are important."	Female, Study Investigator
	"Although it's hard to achieve, it's best not to have a rotating door study coordinators but to try to have the same study coordinator because I can tell you my patients developed relationships with me, my study coordinator, with my nurse, with these positions being fixed during the study period."	Male, Study Investigator
	"But I do think it depends on the level of experience of the staff you have, right? If you're someone who is very, very new, it would be a colossal mistake to put them to do a study like this, because if we think about issues of trust, of science, you can break that trust in many different ways. And one of those ways is to have a bad experience and participate in research."	Male, Study Investigator

TABLE C-2h Attaining Resources and Support to Accomplish a Representativeness

Subthemes	Selected Quotes	Gender/Role
Inclusiveness should be a national priority	"Inclusiveness in research should be a national priority. Again, for the reasons that not only do we need science that provides us data relevant to our population so we're not just studying while men and using the information to treat black men or black women or vice versa. That to me is really, really important and should be a national priority."	Male, Study Investigator
	"But even with the clinical and the pharmaceutical companies, they are going to benefit because they're going to prescribe these medications to everybody. And does it work for everybody? I don't know. Do we know? And they don't need to. They don't. There's no pressure on them to find out."	Female, Study Investigator
Funding needs to be increased for studies that prioritize inclusiveness	"You have to have specifically motivated program announcements . . . towards communities that have been traditionally left out."	Female, Study Investigator
	"So you couldn't have done this study on a typical R01 and the biggest reason for that is the extra cost."	Female, Study Investigator
	"I think that it would be good for efforts to recruit and retain these folks, to have potential additional budgeting so like it's a $500,000 grant but you're going to recruit over 40 percent folks with lower socioeconomic status, than there's an extra $50,000 a year for direct costs to support those efforts. I think we have to put our money where our mouth is, and I don't see that is happening. Especially because what is happening is that you're being held to task a lot more as a clinical trialist, you're held to task a lot more for hitting your recruitment targets. And so an acknowledgment that I can easily hit those recruitment targets with the wrong people. But I cannot easily hit them if I'm being really intentional about this, so we need some sort of incentive to balance that."	Female, Study Investigator
	"You have to include the community partners and the community organizations staffing and what they need. You have to include a timeline that also takes into consideration because it does take much longer to establish and start up a study in the community setting because they have other priorities. They have a lot of other stuff going on. . . . It's not realistic to expect it to be something that fits the model of what's done in the academic ivory towers. So, yea, it's really budgeting for time and effort of people who are not typically thought of in grants. But those are like most of my grants, like all of the funding goes externally . . . like subcontracts to different partners."	Female, Study Investigator

APPENDIX C

TABLE C-2h Continued

Subthemes	Selected Quotes	Gender/Role
	"Research funding has been so disproportionately low in these communities compared to the population size and how we need to rewrite these rules. We need to put directed funding into these types of research."	Female, Study Investigator
	"We really did want at least a partial rural sample because that's probably who needs telehealth the most. And so we did have the funds to do that driving. We had, it was a large R01 trial. So, I mean our budget was hardy and we were able to have a research coordinator, which was me, do the driving and I was just part of the job . . . That as the work time of driving that was certainly part of that 40 hour work week."	Female, Study Investigator
	"The budgeting is really critical. You mention that because if you don't have enough money to hire a recruiter, you've just kept your budget for personnel really tiny in your head. So I see this in a lot of studies I review. So there's many people who are Co-Investigators on these projects and they're all 10% effort. But you have very few people on the ground doing the work. You're not going to be successful, certainly with this more difficult group to recruit, because it takes a lot of time."	Female, Study Investigator
	"We need more personnel than expected. We have to hire more RAs than we thought. And it wasn't really expensive. And I know since then NIH has done some efforts to make the R01 four years. And it wouldn't have been possible in that scenario. So I do think financially finances are a big thing."	Female, Study Investigator
	"I guess I would just say that I think it's a combination of both a site-level responsibility and also a sponsor-level responsibility. I think it would be much harder to accomplish and increase diversity in clinical trials if both parties weren't doing everything they could to try and improve diversity."	Female, Study Coordinator
	"I think everything comes back to money and time. You know, I think the key is having time to make those connections, having time to reach these communities. But time equals money. And so I think it's just understanding that it's a priority. And so I do think particularly with NIH, each year they cut our budget and then we have to reallocate time and figure out how that's going to work and all that. But it's nice to be able to have a cushion for that infrastructure to be able to do these things."	Female, Study Investigator
	"But, unfortunately, as you know, a lot of our budgets don't allow us to do it. We're barely making it through the study and we're like, oh my goodness. But I do think funding agencies need to pay attention to . . . how we have that sort of close relationship right with the participant."	Male, Study Investigator

continued

TABLE C-2h Continued

Subthemes	Selected Quotes	Gender/Role
	"Let's say I budget myself for 20% on the time on a grant knowing that 5% of that time I'm going to be using that for outreach. Then when, each year, I can build that into my grant when I write that. But then when they cut my budget the next year, then I go down to 15% because that's what it's budgeted for. The university gives us a raise but NIH says no raises and so then your time has to go down. And so even if I plan ahead and I build in this time and it keeps getting less and less, then it's hard to figure out where that comes from."	Female, Study Investigator
	"I would say they could look to fund protocols or projects that are written by people of color, not white physicians at large academic institutions. So look to fund more diverse PI populations."	Female, Study Coordinator
	"I think that for me one thing that would help is if funding agencies provided additional funding, there was an option for supplemental funding, for example, to bring on sites that could enhance the diversity of the population."	Female, Study Investigator
	"You have grants that are specifically for underrepresented minorities and their application process is different. . . . . NIH just had these transformational applied research grants, [they] were the first time that they ever had them. And so I think something like that is getting there but it's still too academic. . . . We are scientists. We want to ask research questions but the reality is it misses this really large group of very smart people doing really good work and we're missing it. We're not funding it."	Female, Study Investigator
Provide material resources to community organizations	"It's also bringing resources to them. So we're having grants to fund them and their staff and the work they do and the other programs that they care about building and working on. So having, working that into grants, it's really important."	Female, Study Investigator
	"I think that they actually have to provide financial incentives for organizations, actually nursing staff, as well as the organizations to care about these issues. I mean, really facilities do care about their operational and management issues. Those are key priorities for the resources, not environment at all, even though they may have a great future. And there is some movement to create, and I'm talking about typically financial or physical material. . . . So resources, concrete resources, financial incentives to our facilities that actually care. That give up more of a diverse representation . . . . I think that there has to be some kind of flexibility in terms of how we can use the fund to actually motivate and engage facilities and providers. They are huge gatekeepers."	Female, Study Investigator

TABLE C-2h Continued

Subthemes	Selected Quotes	Gender/Role
	"Also look to fund research centers . . . in more rural populations, or in cities that have a larger Black population, larger Hispanic population. Yea, I would say throw money at the diverse population."	Female, Study Coordinator
Creating education resources and networks for study teams at federal and institutional level	"I think education, I mean recruiting more leaders and then education for those leaders, all in the spectrum in health care."	Female, Study Investigator
	"This original trial we started recruiting back in XXXX. And, since then, I have really seen an increase in recruitment strategies at national presentations we go to. There's been a lot more preconference workshops on it and just presentations and general symposiums. . . . So I think education has certainly helped. I think that's been a good first step."	Female, Study Investigator
	"I think that I attend diversity trainings as voluntary once a month and we have forums and different speakers and it's just opened my eyes up so much and made me very sensitive to how important all this is."	Female, Study Investigator
	"I've always thought that our universities, bit as it really is, why isn't there an office? I mean, there's an office of diversity, which as you know, they organize all those panels that I go to and people with disabilities, it's all different. So it gives you this whole picture of all different kinds. I just don't understand why and I tried to talk to somebody here and she was not helpful at all, but it should be. How about a group of underrepresented people that are paid to do this, to have connections in the community so that you trust her . . . Somebody that already has these connections so that they can help with cancer research, recruitment, or Alzheimer's or anything else. I mean, why isn't it? Seems like a good way to go and I just feel like universities haven't really done that. These little committees I go to, people say, oh, you might be able to call this but there's just, they don't seem to have a real super interest in it."	Female, Study Investigator
	"Put them in consortiums of them together and then try to see what we find in this community. And why can't they do that match? . . . There's a lot of people studying Asian here and we're all disconnected. Big no-no, NIH should know everyone with their funding for Asian studies. Why don't they also say, 'Hey guys, work together. We're giving you millions of dollars each to do this. What are you all doing together?'"	Female, Study Coordinator
	"I do think that federal government and funding agencies need to invest money into looking for best practices. Not just in one place, with best practices that can be scalable to various settings because, yes, we need to enroll trials in high diverse areas but we need to also get people in on diverse areas to learn how to be more effective at recruiting diverse patients in the clinical trial."	Male, Study Investigator